Paul Carus

Homilies of Science

Paul Carus

Homilies of Science

ISBN/EAN: 9783337035808

Printed in Europe, USA, Canada, Australia, Japan

Cover: Foto ©berggeist007 / pixelio.de

More available books at **www.hansebooks.com**

HOMILIES

OF

SCIENCE

BY

DR. PAUL CARUS.

πάντα δὲ δοκιμάζετε,
τὸ καλὸν κατέχετε.
      Paulus ad Thess. I Ep. 5, 21

CHICAGO:
THE OPEN COURT PUBLISHING COMPANY.
1892.

TO
The Sacred Memory
OF
My Father
THE LATE
Gustav Carus
First Superintendent General of the Church of Eastern and Western Prussia and Doctor of Theology

Who would not have agreed to the main doctrines of this Book

But whose life exemplified its teachings.

PREFACE.

THESE HOMILIES OF SCIENCE first appeared as editorial articles in *The Open Court*. The principle that pervades them is to preach an ethics that is based upon truth and upon truth alone. Truth is a correct statement of fact. Truth accordingly is demonstrable by the usual methods of science, and whenever a statement appears to be incorrect or insufficient everybody has a right to examine it, either for refutation or verification, and in this sense the book was named "Homilies of Science."

There is a difficulty in writing Homilies of Science. This difficulty consists mainly in the fact that they must appeal through thought to the will; they must convey sentiment without being sentimental; they should not employ emotional arguments and they have to dispense with all the charms of traditional religious poetry. Moreover they stand in opposition to and have to counteract a very popular error, viz., the view that a full knowledge of the laws of this world would rather dispose a man to become immoral than to purify and ennoble his soul. The belief is not uncommon that a moral teacher has either to suppress some of the facts or to add some fictitious facts. The rules of morality it is often supposed, can be justified through pious fraud alone.

If that were so, morality would stand in contradiction to science and the holiest feelings, the deepest wants, the highest aspirations of mankind would be mere illusions.

The "Homilies of Science" are not hostile towards the established religions of traditional growth. They are hostile towards the dogmatic conception only of these religions. Nor are they hostile towards freethought. Standing upon the principle of avowing such truths alone as can be proved by science, they reject that kind of freethought only which refuses to recognise the authority of the moral law.

The Religion upheld in these Homilies may be called Natural Religion to the extent that it takes its stand upon the facts of nature, that is, the experiences of life or the data furnished us by the world in which we live. It may be called the Religion of Science in so far as the statement of these facts must be done with scientific exactness and critical circumspection. It may be called the Religion of Humanity, in so far as it finds its aim in the elevation, progress, and amelioration of mankind. It may be called Cosmic Religion in so far as its ethics rests upon the consideration that every individual is a part of the great whole of All-existence. It may be called the Religion of Life, for it is concerned with the salvation of the human soul, so as to make man fit to live and to meet the duties of life. Or it may be called the Religion of Immortality, for it teaches us how through obedience to the moral law our lives can become building-stones in the temple of humanity which will remain forever as living presences in future generations. It preserves the human soul, even though the body die, and gives it life everlasting.

* * *

Many a reader will ask, how did this peculiar combination of seemingly opposed ideas come about which are on the one hand so unflinchingly radical and iconoclastic and on the other hand so tenaciously conservative and religious ? The answer is, They developed naturally; they are the result of the author's life, and the product of his experiences.

From my childhood I was devout and pious, my faith was as confident as that of Simon, whom, for his firmness, Christ called the rock of his church. On growing up, I decided to devote myself as a missionary to the service of Christianity. But alas! inquiring into the foundations of that fortress which I was going to defend, I found the whole of the building undermined. I grew unbelieving and an enemy to Christianity. Yet in the depth of my soul I remained thoroughly religious. I aroused myself and gathered the fragments from the wreck, which my heart had suffered. Instinctively I felt that some golden grain must be amongst the chaff.

When my confidence in dogmatic Christianity broke down, I lamented the loss, but after I had worked my way through to clearness I saw that the pure gold is so much more valuable than the ore from which it is gained. I have lost the dross only, the slags and ashes, but my religious ideals have been purified. My life was such that I could not help becoming a missionary, but I became a missionary of that religion which knows of no dogmas, which can never come in conflict with science, which is based on simple and demonstrable truth. This religion is not in conflict with Christianity. Nor is it in conflict with Judaism or Mohammedanism, or Buddhism, or any other religion. For it is the goal and aim of all religions.

I see now Christianity, and the other religions also, in another light. The old Christianity had to stand or fall with certain dogmas. The new Christianity is identical with truth. It is no longer belief in a dead letter, but faith, a living faith in truth; and no scientific progress will ever destroy it.

Every religion has the tendency to drop all sectarianism and to develop into broad humanitarianism. Every religion will in its natural growth mature into a cosmical religion.

How many thousand hearts investigate like me! They have believed and doubted, they have criticised and condemned. And how many that winnow the wheat, lose the grain together with the chaff!

* * *

I hope that wherever my work is inadequate, and I heartily wish it were better in every respect, others will come after me to do it better than I did.

THE AUTHOR.

"Prove all things; hold fast that which is good."—*St. Paul.*

"Free enquiry into truth from all points of view is the sole remedy against illusions and errors of any kind."—*Herder.*

"Let there be no compulsion in religion."—*Koran.*

"The ink of the sage and the blood of the martyr have the same value in heaven."—*Koran.*

"Ein Mensch ohne Wissenschaft ist wie ein Soldat ohne Degen, wie ein Acker ohne Regen; er ist wie ein Wagen ohne Räder, wie ein Schreiber ohne Feder; Gott selbst mag die Eselsköpf nicht leiden."—*Abraham a Sancta Clara.*

> "Science ye shall honor
> Far from vainglorious pride.
> For God's are those who teach,
> And God's are those who aspire.
> He who science praises, praises God."—*Koran.*

"God is an empty tablet upon which nothing is found but what thou hast written thyself."—*M. Luther.*

"Despair alone is genuine atheism."—*J. Paul Richter.*

"God is wherever right is done."—*Schiller*

"The purpose of true religion should be to impress in the soul the principles of morality. I cannot conceive how it has come about that men, especially the teachers of religion, could deviate so far from that purpose."—*Leibnitz.*

The world order is the basis of ethics.

TABLE OF CONTENTS.

RELIGION AND RELIGIOUS GROWTH.

	PAGE.
Is Religion Dead?	1
To Fulfil Not to Destroy	4
The Vocation	8
Religion Based upon Facts	13
The Religious Problem	18
New Wine in Old Bottles	22
The Revision of a Creed	28
The Religion of Progress	32

PROGRESS AND RELIGIOUS LIFE.

The Test of Progress	36
The Ethics of Evolution	43
Fairy-Tales and Their Importance	48
The Value of Mysticism	52
The Unity of Truth	58
Living the Truth	63
Thanksgiving-Day	68
Christmas	71

GOD AND WORLD.

Revelation	75
God	79
Design in Nature	83
The Conceptions of God	90
Is God a Mind?	99
Is the Infinite a Religious Idea?	108
God, Freedom, and Immortality	113
Prometheus and the Fate of Zeus	117

THE SOUL AND THE LAWS OF SOUL-LIFE.

	PAGE.
Enter Into Nirvana	121
The Human Soul	127
The Unity of the Soul	133
Ghosts	137
The Religion of Resignation	143
The Religion of Joy	148
The Festival of Resurrection	151

DEATH AND IMMORTALITY.

The Conquest of Death	155
The Price of Eternal Youth	158
Religion and Immortality	163
Spiritism and Immortality	166
Immortality and Science	174
Death, Love, Immortality	185

FREETHOUGHT, DOUBT, AND FAITH.

Freethought, its Truth and its Error	189
The Liberal's Folly	195
The Mote and the Beam	200
Superstition in Religion and Science	206
The Question of Agnosticism	213
The Bible and Freethought	221
Faith and Doubt	227
The Heroes of Freethought	250

ETHICS AND PRACTICAL LIFE.

The Hunger After Righteousness	233
Ethics and the Struggle for Life	239
Render Not Evil for Evil	245
Religion and Ethics	252
The Ethics of Literary Discussion	256
Sexual Ethics	260
Monogamy and Free Love	263
Morality and Virtue	269

SOCIETY AND POLITICS.

Aristocratomania	276
Socialism and Anarchism	283
Looking Forward	288
Womon Emancipation	294
Do We Want a Revolution	298
The American Ideal	305

HOMILIES OF SCIENCE

IS RELIGION DEAD?

ONE of the greatest historians of morals says: Religion has ceased to be the moving power in our national and in our private life. Interest in theological discussions is nowhere to be found, not even in the churches. What do the people care for the religious issues of former days? They are quite indifferent about the interpretation of Bible passages and the sacraments, which in former centuries caused sanguinary wars among nations. And a great French philosopher announces the advent of an irreligious age, where creeds will disappear, where no church shall exist, and religion shall cease to be.

Contemplating the habits and the life of our age, we are struck by a noticeable change in the general tendencies of men. It seems that everything has become more worldly, more realistic, and more practical. Yes, more practical! and I should say there is no harm in being practical, if the ideal world be not lost in the realistic aims which we pursue, if our hearts be still aglow with the sacred fire of holy aspiration for purity, for honor, and above all, for truth! Let us be practical, and let us more and more become so, in applying the highest ideals to our everyday life and in realizing them!

The God of old Religion said through the mouth of one of his prophets: "Lo, I make all things new." And a psalmist of the western world sings in one of his deepest lays: "There is no death—what seems so, is transition." Nature cannot die, it may undergo changes, but it will live forever. Nature is life, it is the fountain of eternal youth.

Learn to understand the signs of the time. If you see the leaves turn yellow and red and shine in all colors, know that autumn is at hand. The leaves will fall to the ground and snow will soon cover the trees and woodlands and meadows. But when you see buds on the branches, although they may be few and the weather may be cold, still, know that spring is at the door, and will enter soon, filling our homes with flowers, with joyous life, and with love.

The leaves of dogmatic opinion are falling thickly to the ground. How dreary looks the landscape, how bleak the sky! How cold and frosty, how forlorn are the folds of the churches! There is the end of religious life, you think; the future will be empty irreligiosity—without faith in the higher purposes of life, without ideals to warm and fill our hearts, without hope for anything except the material enjoyments of the present life.

And yet, my friends, observe the signs of the time! There are buds on the dry branches of religious life which show that the sap is stirring in the roots of the tree of humanity. There are signs that the death-knell of the old creeds forebodes the rise of a new religion.

Everyone who knows that nature is immortal can see and feel it. A new religion is growing in the hearts of men. The new religion will either develop from the old creeds which now stand leafless and without fruit, which seem useless, as if dead, or it will rise from the very opposition against the old creeds, from that opposition which is made not in the name of frivolous cynicism, but in the name of honesty and truth. The beautiful will not be destroyed together with the fantastic, nor the higher aspirations in life with supernatural errors. Though all the creeds may crumble away, the living faith in ideals will last forever. That which is good and true and pure, will remain—for that is eternal.

The new religion which I see arising and which I know will spring forth as spontaneously and powerfully as the verdure of spring, will be the religion of humanity. It will be the embodiment of all that is sacred and pure and elevating. It will be realistic, for it loves truth. It will promote righteousness, for it demands justice. It will ennoble human life, for it represents harmony and beauty.

The new religion that will replace the old creeds will be an ethical religion. And truly all the vital questions of the day are at bottom religious, all are ethical. They cannot be solved unless we dig down to their roots, which are buried in the deepest depths of our hearts—in the realm of religious aspirations.

Life would not be worth living if it were limited merely to the satisfaction of our physical wants; if it were bare of all higher aspirations, if we could not fill our soul with a divine enthusiasm for objects that are greater than our individual existence. We must be able to look beyond the narrowness of our personal affairs. Our hopes and interests must be broader than life's short span; they must not be kept within the bounds of egotism, or we shall never feel the thrill of a higher life. For what is religion but the growth into the realm of a higher life? And what would the physical life be without religion?

TO FULFIL NOT TO DESTROY.

The greatest religious revolution which the world has ever seen was that of Christianity. From the standpoint of an impartial umpire, it must be confessed that the triumph of the Christian Faith has been the grandest in history. The founder of Christianity, who died on the cross as an outlawed criminal, led the van of a new civilization. In his name kings and emperors reverently bowed and yielded to the demands of humaner ideals; while the greatest philosophers, the princes of thought, brooded over his ethical doctrines.

How can we explain the unparalleled success of Christianity? It is due, undoubtedly, to the sublimity of Christ's ethics, to the gentleness and nobility of his person, to the kindness of his heart, to the wealth of his spiritual treasures, and to the poverty of his appearance. But that is not all. Every business man knows that for success, not only ability is required, not only the solidity of one's goods, but the merchandise offered must also be in demand.

No movement in history can be successful unless it is based upon a solid ethical basis, having in view the elevation and amelioration, not of a single class or nation, but of the human kind. Yet this is not all. A revolution must be needed; it must stand in demand. No revolution will endure unless the ethical idea by which it is animated lies deeply rooted in the past.

A successful *revolution* must be the result of *evolution ;* and a successful revolutionist must combine two rare qualities, an unflinching radicalism and a strong conservativism. The ideal of a successful movement must open new and grand vistas for progress, but at the same time it must be the fulfillment of a hope, the realization of a prophecy. Thus it will shed its light on the ages past, which will now be understood as preliminary and preparatory endeavors to effect and to realize this ideal.

We stand on the eve of another great religious revolution. Humanity has outgrown the old dogmatism of the churches, and a new faith is bursting forth in the hearts of men, which promises to be broader and humaner than the narrow bigotry of old creeds. It promises to accord with science, for it is the very outcome of science ! It will teach men a new ethics —an ethics not founded on the authority of a power foreign to humanity, but upon nature, upon the basis from which humanity grew ; it will rest upon a more correct understanding of man and man's natural tendency to progress and to raise himself to a higher plane of work, and to a nobler activity.

Science has undermined our religious belief, and beneath its critical investigations dogmas crumble away. But whatever science may undermine of ecclesiastical creeds, it does not, and will not, prove subversive of the moral commandments of religion. Science will, after all, only purify the religious ideals of mankind, and will show them in their moral importance. The most radical criticism of science will always remain in concord with the reverent regard for the moral ideal.

We believe in progress, and trust that man lives not in vain, that man's labor, if rightly done, will further the cause of humanity and make the world better —be it ever so little better—than it was. We aspire to a nobler future—and let me point out one important subject which is too often overlooked, and which is indispensable to success. The success of ideals is impossible without a due respect for the ideas which are to be displaced. The triumph of a better future depends upon a due reverence for the merits of the past, or, in other words, we must know that the new view is the outcome of the old view. The ethical religion of the future springs from the seed of past ecclesiastical religions. And if the latter appear to us as superstitious notions of a crude and strangely materialistic imagination, they nevertheless contain the germs of purer and more spiritual conceptions. And there is no doubt that the founder of Christianity is more in accord with the new rising movement than with the doctrines of his followers, who worship his name, but neglect the truth and spirit of his teachings.

When Christ preached the sermon on the mount, which contains, so to say, the programme of his doctrines, he expressly stated: "Think not that I am come to destroy the law or the prophets; I am not come to destroy, but to fulfil." This sentence contains the clue to his grand success. Christ was a conservative revolutionist. The new movement which he introduced in the history of mankind, was the result of the past; the New Testament was the fulfillment of the Old. And so every successful movement has been, not a mere destruction of old errors, not the introduction of some absolutely new idea, but the fulfillment of the past, and the realization of long cherished aspirations and hopes.

Let us learn a lesson from Christ, and like him, let us "not come to destroy, but to fulfil."

THE VOCATION.

When I was a youth a voice came unto me and said: "Preach!" And I answered: "What shall I preach? Lo, I am young and have not sufficient knowledge." "Go into the world," I was told, "and preach the truth."

That voice came from my parents and grandparents, from my teachers and instructors; and it found a ready response in my soul. To be a preacher of Truth, what a great calling! Is there any profession more glorious, is there any work more celestial and divine? I will go and preach the truth, I avowed; and in the secretness of my heart I swore allegiance to the Banner of Truth. I vowed to seek for Truth, to find it, to confess it, to go into the wide world and to preach it, yea, to give not only all my labor and efforts, but, if it were necessary, even my life, my blood, myself, and all that I was, for truth.

That was a holy hour in which I devoted myself to the cause of truth, and yet it was a rash decision, a preposterous act. It was an act that I had to regret in many dreary hours when I desperately pondered upon the problems of truth, when I had hopelessly lost myself in the labyrinths of life, and when I despaired of Truth's very existence.

When I was young, Truth seemed so simple to me. What is Truth? I asked, and the teaching of my childhood always echoed forth the ready answer: Truth is the gospel, and doubt in Truth is the root of all evil.

I knew the gospel by heart, and I studied eagerly, that I might be a worthy minister of the word of God. But the more I studied the more that sinful tendency to doubt grew, first secretly, then openly, first suppressed, then frankly acknowledged, until doubt ceased to be doubt; it became an established conviction. A cry of despair wrung itself from my heart: "The gospel is not truth; it is error! It is a falsity to preach it, and he who preaches it, preaches a lie!"

A pang of discord vibrated through my bosom and tore my whole being into two irreconcilable parts. Could I step to the altar in this condition and swear to preach the gospel? Never! I had believed that the gospel was but another name for truth and I now saw that whatever truth might be, the gospel certainly could not be truth.

Is there truth at all? No! I thought; there is no truth! There are opinions only, and one opinion is as good as another. Man likes to look upon the world as a cosmos—but there is no cosmic order, there is no higher law, there is no justice and no truth in the world, there is disorder everywhere, the universe is a chaos of forces, natural laws are indifferent to good or evil, and the lie rules supreme in society, sham gains the victory over truth, cunning and selfishness triumph over virtue and love.

Oh! these were dreary hours when I had lost the ideals of my childhood. I had cast my anchor into the ground of religious belief and had suffered a shipwreck, in which I expected to perish.

There was a time when I did not know which I hated more, Science that had taken away the comfort of my religious faith, or Religion that had promised all to me and had proved false. Religion could not justify itself as Truth before the court of scientific research.

I abandoned religion and followed science.

Years passed away amid earnest labors, and science reluctantly opened to me her treasures. She made me see the wonders of life. Life appeared different to me. The universe of science is another world than that which I imagined to see around me in the chaotic turmoil of the struggle for existence. I perceived invisible threads that connected distant events. I recognized that while the laws of nature might work blindly, yet they produced order. The more my views expanded, the clearer I saw that the chaotic attaches to the single, to the isolated only, not to the whole, not to the greater system, and the All itself is identical with order. The All is a cosmos truly.

Opinions clash with opinions in the empire of science; and the knowledge that we possess is almost always an approximate statement only of the truth. Nevertheless, there is truth and there is error. One opinion is by no means equivalent to every other opinion; there are wrong opinions and correct opinions, there is Truth in this world and Truth is a power. She reveals her sacred face only to him who earnestly struggles for truth. Truth may seem awful at first, but fear her not; trust her, have confidence in her, even as does a child in its mother. Give up your prejudices and your misconceptions even if they are holy to you, even if they seem to constitute the very life-blood of your spiritual being.

In the meantime I had given up every intention to preach the gospel and found satisfaction in the retired hermitage of the study, where I became an adept of truth in quite another sense than I had intended in the preposterous ambition of my youth. I was not a teacher, not a preacher of truth, but her pupil, not a master but a disciple who plodded modestly and patiently. How often, O how often, was a grain of truth dearly bought through the toil of many, many hours—and yet never too dearly!

In former years I had answered the question What is truth, with the words: "Truth is the gospel." Now I learned to reverse the statement. I had met so much misery and woe in the world and in looking around for salvation, I said: If there is any gospel, it must be truth—and truth must be found by patient labor, by scientific, honest research and by severe exactness. What a folly in man to imagine that truth should drop down from heaven as a revelation. Truth must be conquered by our own efforts. Truth would not be truth if it were acquired in some other way.

Years passed away and, again a voice come unto me and spoke: "Preach! Preach the truth." I answered and said: "How can I preach? Am I not a mere disciple who has no hope ever to become a master? I am no preacher and no one has appointed me to speak in the name of truth. When I was a youth I felt the strength to preach, and lo, I had it not. I had almost stepped to the altar and had almost made a vow which I now know I should have had to break. Let me study truth, let me devote my labor to science, but send another man worthier than I. Besides, I am not eloquent: but I am slow of speech and of a slow tongue. I know not how to speak as a preacher to the congregation.

But that voice came again: "Preach the truth." He who is called to proclaim the religion of mankind will not be bound by any oath to adhere to this or to that confession of faith. He is pledged to be faithful to truth only. If you have the conviction that truth—mere truth and nothing but the truth—will be the gospel of mankind, that the salvation from error can come from it alone, that science, whose fruit seemed so bitter at first, contains the germs of a higher religion, step forward upon this platform and preach that new faith which is greater than the old faith, because it is truer.

I feel as if a preacher that has not joined any of the many churches, must be a voice crying in the wilderness. But that should be no reason to decline the calling. Therefore I shall accept the call upon that platform. One thing alone shall be sacred to the preacher of the religion of humanity, and that is truth. There shall be no oath of allegiance to any dogma, no pledge to any creed. I accept the calling, yet I do it with hesitation, because I am aware of its difficulties. And at the same time I accept it in gladness, because I know that the new religion which grows out of science—out of the rock upon which the old creeds were shipwrecked— will not come to destroy. The new religion will come to fulfill the old faith.

RELIGION BASED UPON FACTS.

A well known clergyman, famous for his indefatigable energy, and the comprehensiveness of his practical activity, who believed in a supernatural world of purely spiritual existence, and a scientist with materialistic tendencies who looked upon all religious aspirations as mere illusions, once had a discussion about facts. The scientist declared that science alone dealt with facts, the clergy did not see the real world, but dealt with things that were unreal. The clergyman answered rather sharply in about this way: "You scientists imagine that you have a monopoly of facts. You should know that I have to deal with facts just as much as you do. I have stood at the bed-side of the sick and dying, and my experiences concerning that which comforts them in the hour of death and tribulation are based upon observations of facts. Practical theology is in no less a degree based upon facts than the science of physical or chemical phenomena."

The clergyman was right in so far as the duties of his calling arose from the facts of life. A pastor should be the adviser, the fatherly friend, and comforter of his congregation in all the situations of life. Individuals are not isolated beings. Many of their actions, and indeed their whole demeanor, are of great concern to the community, and the community protects itself against vicious individuals by law. The duty of the clergy is to impress upon their congregations the moral spirit of goodwill towards all mankind, to teach them to regulate their conduct so that in the hour of death no remorse will flit over their minds,—to teach them that when they lie down to eternal rest, their deeds, their love, their sympathy, and their thoughts will live on and bear witness to their having fought a noble battle in life. The more thoroughly the clergyman does his duty in a spirit of religious truth and moral aspiration, the less will we want the work of the state's-attorney and the judge.

It is to be hoped that our churches will imbibe more and more the positive spirit of the age, and so found their duties upon the facts of life. Whether they believe in a supernatural world of purely spiritual existence is, or should be, of secondary importance. Our churches, however, have so much mixed up the real and objective facts of life with their antiquated interpretations of these facts, that they believe the fictitious world of supernaturalism as described in their dogmas, to be a reality.

It is a fact that people need solace in the hour of death, it is a fact that matrimony is a holy ordinance, in which not only the couple that is united for life until death do them part, but the whole community is greatly concerned. It is a fact that the birth of a child imposes duties upon the parents; the child is not their property; it is entrusted to their care, and they have to rear it for the best of humanity. Godfathers or godmothers promise to take the place of parents, if death should call the latter away too early to fulfill their duties upon the child. From the naming of a child upon its entrance into the world, unto the burial of the dead, when we pay the last honors to our beloved ones, man's life is permeated with duties that point higher than the fulfillment of egotistic desires. Egotism finds its end in death; man's duties teach him to think beyond his own death. And it is the performance of these duties that is the substance of all religious commands.

Some imagine that science is limited to the lower sorts of natural facts only. Religious and moral facts have been too little heeded by our scientists. Thus people came to think that science and religion move in two different spheres. That is not so. The facts of our soul-life must be investigated and stated with scientific accuracy, and our clergy should be taught to purify religion with the criticism of scientific methods. They need not fear for their religious ideals. So far as they are true, and their moral kernel is true, they will not suffer in the crucible of science. Religion will not lose one iota of its grandeur, if it is based upon a scientific foundation; all that it will lose is the errors that are connected with religion; and the sooner they are lost the better for us.

One of my orthodox friends maintains that Christianity, that is to say orthodox Christianity, is based upon facts, and these facts, he says, are historical facts: they are the life and teachings, the suffering and the death, and above all the resurrection, of Jesus Christ.

If Christianity is based solely upon historical facts, it stands and falls with their truth. If Christian morals depend upon the occurrence of a few events that are supposed to have happened once and will never happen again, their fate is very problematic indeed.

The question is well worth a closer consideration.

Natural processes around us show a certain regularity combined with a certain irregularity. Every phenomenon that takes place has its individual features, and no one thing is exactly like another. A visitor from the city may imagine that every sheep in a herd of one breeding looks like the other; yet the shepherd knows them all individually, and can distinguish them apart. Grains of corn may appear to us all alike, yet they are not; every one has its own idiosyncrasy. But in spite of all difference, there is a universality of law in all things and in all natural phenomena. A closer acquaintance with the nature of the differences teaches that they result, and can only result, from a difference of condition. Yet it is the same law that governs all. Thus we arrive at the conclusion, that isolated facts cannot exist which stand in contradiction to the laws of all other facts. And it is a rule that science derives its laws—the so-called natural laws—from such facts alone as repeat themselves again and again, from such as can be verified by experiment, from such as are accessible to the observation of every one who takes the trouble to investigate. It need scarcely be added that the same rule holds good for positive philosophy. Single and isolated observations cannot give a solid basis for a conception of the world. The facts upon which a view of the universe rests must be ascertainable by every one who cares to be positive about their being as they are represented to be and not otherwise.

The rule is unequivocally acknowledged in science. It is accepted—by some with a certain reserve—in philosophy. Yet it is recognized in religion only by few. Although if it be true in science it must be true in religion also.

What is religion but a conception of the world, in accordance with which we regulate our conduct? If religion is based upon verifiable facts, it stands upon a rock. If it is based upon an assertion of facts that happened once and will never happen again, it is built upon sand; and when 'the rain descends, and the floods come, and the wind blows, and beat upon it,' the structure will fall.

Christ's doctrine in so far as it is the religion of love, stands upon the moral facts of human soul-life. The ethical truth of Christianity rests on solid ground. Christian dogmatism, however, stands or falls with the history of Christ's life, his death, and resurrection. Had not orthodox Christianity been supported by the great truth of Christ's religion of love, it long ago would have disappeared; for Christianity as an historical religion is indeed extremely weak. What must a religious truth be that has to depend upon the verification of a few historical facts? And these historical facts are in themselves improbable, nay, impossible; they stand in contradiction to all the facts verified by science, and whether they are true or not, have not the least bearing upon the moral conduct of man. Whether Christ healed a few lepers or not, whether he abstained from all food for forty days or not, whether he has bodily risen from the dead or not, the 'ought' of Ethics remains the same. If Christianity means the dogmatism of the Church, it is an historical religion which will disappear in the course of time; if it means the doctrine of Christ, the fulfillment of the law through love, it will be the religion of mankind.

THE RELIGIOUS PROBLEM.

The political, religious, and intellectual growth of humanity constantly produces changes in the conditions of society, and in times of rapid progress these changes may become so great as to demand the readjustment of our institutions of government, the reformation of church and school, and the reconstruction of our fundamental conceptions of the world and life. When the necessity, therefore, for readjustment and reformation becomes keenly felt, problems arise. Thus we speak of the social problem, the educational problem, the religious problem, and many others.

The religious problem results from the rapid advances made by science. Our religious conceptions, it is now generally acknowledged, can possess value only if they are recognized in their moral importance. Their dogmatic features are coming more and more to be considered as accessory elements, which can, and indeed often do, become injurious to the properly religious spirit.

The moral rules which we accept as our maxims of conduct in life, must have some basis to rest upon. We demand to know why and to what end the single individual has to obey certain commands, to observe which may sometimes cost great self-sacrifice. The old orthodox systems of religion cannot answer this question at the present day with the authority which the blind and unasking faith of their adherents formerly attributed to their utterances; and we are therefore brought to the task of remodeling our religious conceptions, in order to make them harmonize with the present altered situation.

The religious problem has been solved differently by men of different stamp. The orthodox theologian, of course, denies the existence of a religious problem. Being stationary he has not progressed with his time; he knows nothing of evolution, and looks upon the advances of science as steps towards depravation. He would solve the problem by checking all further progress, and would keep humanity down to the level of his own littleness.

The iconoclast, on the other hand, solves the problem by extirpating religion altogether. Like Dr. Ironbeard, in the German legend, he frees his patient from pain by a plentiful dose of opium, that lulls him to eternal rest. It is a radical cure. Kill the patient and he will cease to complain.

The religious problem of to-day does not mean that we doubt the ten commandments. We do not object to the behests: "Thou shalt not steal," "Thou shalt not kill," "Thou shalt not bear false witness against thy neighbor." Nor do we object to the Christian ideals of Faith, Hope, and Charity; we do not oppose the rule, "Love thy neighbor as thyself." The religious problem means that we have ceased to believe the dogmas of the church. We have ceased to look upon God as a person who made the world out of nothing, and governs it at his pleasure. We have ceased to believe in miracles; we have ceased to believe in the supernatural and in the fairyland which, according to the dreams of former ages, existed in heaven beyond the skies.

So many illusions fell to the ground when the light of science was thrown upon them; but the moral command, "Love thy neighbor as thyself," did not. Science has destroyed the mythology of religion, but it has left its moral faith intact; indeed, it has justified it; it proves its truth, and places it upon a solid basis, showing it in its simple and yet majestic grandeur.

Science teaches that harmony prevails everywhere, although to our blunted senses it often may be difficult to discover it. Science teaches that truth is one and the same. One truth cannot contradict another truth, and when it seems so it is because we have not found, but will find, the common law that embraces these different aspects of truth which to a superficial inspection appear as contradictory. Science further teaches that the individual is a part of the whole. The individual must conform to the laws of the All, not only to live at all, but also to live well—to live a life that is worth living.

The properly religious truths are not the dogmatic creeds, but the moral commands; and it is their scientific and philosophical justification which is demanded by the religious problem of the present age. The solution of the religious problem must give us a clear and popular conception of the world, based upon the broadest and most indubitable facts of science so arranged that every one can understand the necessity of conforming to those laws which have built human society, and make it possible for us to live as human beings a noble and worthy life. The solution of the religious problem will most likely do away with many sectarian ceremonies and customs, it will enable us to dispense with certain narrow views and antiquated rites, which many, up to this hour, look upon as the essentials of religion. But it will not do away with the moral law; for we know that that will never pass away. It is the moral law which Christ and the Apostles again and again declare contains the essence of all their injunctions: for the whole law is fulfilled in one word, even in this, "Thou shalt love thy neighbor as thyself," and "This is the love of God that we keep his commandments, and his commandments are not grievous."

NEW WINE IN OLD BOTTLES.

CHRIST said : "No man seweth a piece of new cloth on an old garment: else the new piece that filleth it up taketh away from the old and the rent is made worse. And no man putteth new wine into old bottles, else the new wine doth burst the bottles, and the wine is spilled and the bottles will be marred; but new wine must be put into new bottles."

What Christ's meaning was when he spoke these words we can hardly guess, for the context in Matthew (ix, 16, 17) as well as in Mark (ii, 21, 22) appears to be corrupted. Christ, as reported in these passages, said these words in answer to the question: "Why do we and the Pharisees fast oft, but thy disciples fast not?" This part of Christ's answer does not fit to the question. But, whatever Christ meant, it is certain that, if these allegories mean the renewal of old ideas, the rejuvenescence of a dying faith, he himself did pour new wine into old bottles. He did not reject the truths of the Old Testament, but he adopted them, he perfected them, he brought out their moral purport, and showed the spirit of their meaning. If the simile is to be interpreted in this sense, evolution is a perpetual repetition of putting new wine into old bottles.

What is the progress of science but a constant remodeling of our scientific conceptions and terms and formulas? What is the progress of national and social life but a constant alteration and improvement of old institutions and laws?

What enormous changes has our conception of God passed through! How great they are is scarcely apparent to us now, at least our orthodox brethren are not much aware of it. It is known to the historian; and we can give an idea of these changes by pointing to the fact that the idea of evil passed through the same phases. The crude anthropomorphism displayed in the history of the idea of the devil is fresher in our minds, and is better preserved in legends.

How often have the orthodox on the one hand, and infidels on the other, declared that if the word God means anything, it means and can mean only some one thing. How often did the former conclude from such a premise that everyone who did not hold their opinion was an atheist, and the latter maintain that this conception being wrong, there was no God at all. How often was the conception of God changed, and how often had the dogmatic believer to shift his position.

There is a point of strange agreement between the old orthodox believers and their infidel antagonists. Believers, as a rule, declare that religion means nothing, unless it means the worship of a supernatural divine personality; and atheists, accepting the latter definition of religion, conclude that religion, therefore, should be rejected as a superstition.

This agreement between believers and infidels is at first startling. In my childhood I sided with the former, in my youth with the latter; but, when I became a man, I freed myself from the narrowness of both. I now know that some errors they have in common.

Opponents have always something in common, else they could not be antagonistic to one another. Thus the orthodox believer and the infidel disbeliever stand upon the same ground, and this ground is their common error. The infidel speaker on the platform, appears to me, in principle as well as in method, like an inverted orthodox clergyman. He agrees with his adversaries in the principle—and he always falls back upon the dogmatic assertion—that there is no one who can know: no one who can solve the religious problem, no one who can prove or disprove whether there is a God and an immortality of the soul or not. But the infidel inverts the argument of the orthodox believer. While the latter argues, "I must believe, because I cannot know, I must have faith, because it is beyond the ken of human reason;" the infidel concludes, "because I cannot know, I must *not* believe, and I must reject any solution of the problems of God and the soul because the subject is beyond the ken of human reason."

Weighing the pros and the cons of the question, I became convinced that both parties were one-sided, that, misguided by a narrow definition, both had become so ossified as to allow of no evolution to a higher standpoint. Therefore, I discarded all scruples about using the words Religion, God, and Soul in a new sense, which would be in conformity with science. It was, perhaps, a new path that I was traveling, and there are few that find it, but it is, nevertheless, I am fully convinced, the only true way that leadeth unto life.

The adherents of the new religious conception are in the minority; and there are the theists on the one side, and the agnostics on the other, both uniting their objection to a widening of ideas that have become too narrow for us now, both declaring that old definitions should not be used in a new sense.

Strange! is it not? It seems so, but it is not. The agreement between believers and unbelievers is easily explainable from the law of inertia. The law of inertia holds good in the empire of thought just as much as in the empire of matter.

When Lavoisier discovered that fire was a process of oxidation, he met with much opposition among his co-workers. It was plainly told him that fire, if it meant anything, meant a certain substance, scientifically called "phlogiston," the qualities of which could be perceived by our senses. And this phlogiston, it was maintained, possessed, among other properties, the strange property of a negative weight, and the argument seemed so evident, since all flames tend upwards. If fire meant a mere mode of motion, would not that be equivalent of denying the real existence of fire altogether?

We now all know that the definition and the meaning of the words fire and heat have changed. Neither have the words been discarded, nor have we ceased to believe in the real existence of fire, since we have given up our wrong notion of the materiality of fire. On the contrary, we now know better what fire is, and in what consists the reality of a flame.

Concerning religion let us follow the example of Christ, and break the fetters that antiquated definitions impose upon us. Not the letter giveth life, but the spirit; and let us preserve the spirit of religious truth, if need be, at the sacrifice of the letter, in which the spirit is threatened to be choked.

Christ's words about the new cloth, and the new wine, it seems to me, meant that certain religious institutions, that ceremonies and forms will wear out like old garments, and like old bottles. Antiquated institutions, which have lost their sense, should not be preserved. For instance, the sacrifices of lambs and goats, which were offered by the Jews, as well as by the Greeks and the Romans, were abandoned in Christianity: they had lost their meaning, and Christ's religion would have been an old garment with a new piece of cloth on it, if the old cult had been preserved. Indeed, even the Jews are so much imbued with the new spirit that they have given up their sacrifices forever.

It will be the same with the new religion that is now dawning upon mankind. Some of the old ceremonies have lost their meaning, they will have to be dropped. But the whole purport of religion, the ideal of religion and its mission will not be gone. Man will always want a guide in life, a moral teacher and instructor. Man must not allow himself to drift about on the ocean of life, he must have something to regulate his conduct. Who shall do that? Shall man follow his natural impulse to get as much pleasure out of his life as he can? Shall he follow science? Or shall he follow religion?

Man might follow science, if every man could become a scientist; and in some sense, this is possible. We can not, all of us, become specialists in the different sciences, but we can, all of us, to some extent become specialists in ethics. What is religion but a popularized system of ethics? And this religion of ethics will be the religion of the future. All of us who aspire after progress, work for the realization of this religion.

Let the religion of the future be a religion of science, let religion not be in conflict with science, but let the science of moral conduct be so popularized that the simplest mind can obey its behests, not only because he knows that disobedience will ruin him, but also because he has learned to appreciate the moral commands, so as to love them, and follow them because he loves them.

THE REVISION OF A CREED.

We have at present the strange spectacle that in one of our churches the proposition is discussed to change some grave particulars of creed. The old doctrines have become "unpreachable," as it is expressed, either because the ministers no longer believe them, or because people are loath to listen to ideas which now appear as monstrosities and absurdities.

We naturally hail the progress of a church and its development into broader views of religious truth. Yet at the same time we feel the littleness of the advance. What is the progress of a few steps, if a man has to travel hundreds of miles! Moreover, what is any progress, if it is done under the pressure of circumstances only and not from a desire to advance and keep abreast with the true spirit of the times! The change of a creed should not be forced upon a church from without by the progress of unchurched thinkers, but it should result from the growth and expanse of its own life. The church, as the moral instructor of mankind, should not be dragged along behind the triumphant march of humanity, but should deploy in front with the vanguard of science!

The eternal damnation of noble-minded heathen and of the tender-souled infants who happen to die unbaptized, was sternly believed in by the ancestors of our Presbyterian friends. They declared, without giving any reasonable argument for their opinion, that this is part of the divine order of things, and whosoever does not believe it, will be damned for all eternity, together with the wise Socrates and the virtuous Confucius.

Who made Calvin the councillor of divine providence and who gave him the right of electing or rejecting the souls of men? On what ground could his narrow view, excusable in his time, be incorporated into the creed of a church? The argument on which Calvin's view rests, was very weak, but the founders of the Presbyterian Church being convinced of its truth, thought to strengthen it by incorporating the doctrine into their Confession. An idea, once sanctified by tradition, has a tenacious life. Reverence for the founders of a church will keep their errors sacred and will not allow an impartial investigation of their opinions.

Reverence is a good thing; but all reverence toward men, be they ever so venerable, must be controlled by the reverence for truth. And this is the worst part of the change of the Confession. The change, it appears, is not made because the objectionable doctrines are recognized as errors; but simply because they are at the present time too repulsive for popular acceptance.

Why are the doctrines of eternal punishment not openly and confessedly branded as errors? Why can it not be acknowledged that tenets which our fathers considered as truths of divine revelation, were after all their personal and private opinions only?

We ask why, but receive no explanation. Yet there is a reason that lurks behind; although it seems as if the men who are most concerned were not conscious of it. If the error were acknowledged, a principle would be pronounced which opens the door to a greater and more comprehensive reform. And such a reform is not wanted. The clergy seem to be afraid of it. If the error is conceded, it means the denial of the infallibility of the Confession. The dogmas of the church cease to be absolute verities; and truth is recognized above the creed of the church, as the highest court of appeal—truth, *ascertainable by philosophical enquiry and scientific research.*

This would be equivalent to the abolition of all dogmas and wonld mean the enthronement of a principle to fill their place. This principle, if we look at it closely, is nothing new; it is an old acquaintance of ours; it is the same principle on which science stands. And the recognition of this principle would be the conciliation between science and religion once for all.

Brethren, do not shut your eyes in broad daylight, but look freely about and follow the example of the great founder of Christianity. Worship God not in vain repetitions, not in pagan adoration, as if God were a man like ourselves. Worship God in spirit and in truth. Acknowledge the superiority of truth above your creed, and be not ashamed of widening the pale of your churches.

If you acknowledge the supremacy of truth and make your changes in the Confession because truth compels you to make them, your progress will be that of a man who walketh upright and straight. But if you do not acknowledge the superiority of truth above your creed, if you identify truth with your creed, your progress will be the advance of a soldier loitering in the rear of his army, who is afraid of being left behind. You will unwillingly have to yield to the necessity of a change; and you will have to do it again and again, and always without dignity.

Is it dignified to alter a religious creed because it appears as a relic of barbarism, because it has become odious to the people, and because it no longer suits their tastes? Your Confession should be allegiance to truth. Will you degrade it to be the unstable expression of the average opinion of your members?

There is but one way to free yourselves from all these difficulties. Recognize no dogma as absolute and reverence no confession as infallible; but let truth, ascertainable truth, be the supreme judge of all doctrines and of all traditions.

Your bible, your hymn-book, your catechism, the history of your church, and the reminiscences of your venerable leaders shall remain respected among yourself and children, but let them not be overrated in their authority. Truth reigns above them all, and the holiness of truth is the foundation of all true religion.

When Luther stood before the emperor and the representatives of church and state, he begged to be refuted, and if he were refuted, he promised to keep silence; but as he was not, he continued to preach and he preached boldly in the name of truth as one that had authority. Therefore let religious progress be made as in the era of the Reformation, not in complaisance to popular opinion, but squarely in the name of truth.

THE RELIGION OF PROGRESS.

Vladimir Solovieff, a Russian thinker of uncommon depth calls attention to the fact that the central idea of Christianity must be sought in the glad tidings of the kingdom of God. He says:* "To either the direct or indirect elucidation of this idea are devoted almost all the sermons and parables of Christ, his esoteric conversations with the disciples, and finally the prayer to God the Father. From the connection of the texts relating thereto, it is clear, that the evangelical idea of the kingdom is not derived from the concept of divine rule, existing above all things, and attributed to God, conceived as almighty. The kingdom proclaimed by Christ is a thing, advancing, approaching, arriving. Moreover it possesses different sides of its own. It is within us, and likewise reveals itself without; it keeps growing within humanity and the whole world by means of a certain objective, organic process, and it is taken hold of by a spontaneous effort of our own will."

This conception of Christianity is strikingly correct. Taking the gospels of the New Testament as our source

---

*"Christianity: Its Spirit and its Errors." *The Open Court*, Vol. V, No. 206, p. 2900. Translated from the Russian Quarterly *Voprosui Filosofii i Psichologii* by Albert Gunlogsen.

of information, we find none of the Church dogmas proclaimed, but we hear again and again that the kingdom of God is near at hand, and that the kingdom of God cometh not with observation, i. e. with ceremonies or rites. It is not an institution as are synagogues and churches. It exists in the hearts of men. We must create it, we must make it grow within us, Our own efforts are needed to let it come. Says Christ: "From the days of John the Baptist until now the kingdom of heaven suffereth violence, and the violent take it by force."

Is this not a strange conception of the kingdom of God? Indeed it is, if we preserve the orthodox God-idea of a personal world-monarch. But it is not a strange conception of the kingdom of God, if we understand by God the divinity of the universe and the potentiality of spiritual life which has produced us and leads us onward still on the path of progress to ever greater truths and sublimer heights.

What is the meaning of the kingdom of God if we state it in purely scientific terms without using the symbolism of allegorical expressions? God means that reality about us and within us in which we live and move and have our being, and the kingdom of God which has to come, which grows within us, is our knowledge of God, it is our cognition of reality, it is the evolution of truth. What is truth but a correct conception of reality and what is all religion but our agreement with truth in thought as well as in action?

When asked by Pilate whether he was a king Christ said: "Thou sayest that I am a king. To this end I was born, and to this cause I came into the world that I should bear witness unto the truth. Everyone that is of the truth, heareth my voice."

Christ considered himself as a king of truth. "My kingdom," he said, "is not of this world," meaning thereby the world in which the ambition of Pilate was centered. Christ did not intend to exercise political power and the accusations of his enemies as well as the hopes of his followers that he would create a worldly kingdom were unfounded. His kingdom was a spiritual kingdom—the kingdom of truth. Truth however is not something that exists somewhere as objects exist in material reality, truth is the correctness, the validity, the adequateness of our conceptions of reality; and truth does not come to us, we must produce it, we must work it out through our own efforts, we must build it up in our own souls. The more we have acquired of truth, the more we shall partake of the kingdom of God. For Truth is the kingdom of God and the kingdom of God is Truth. Every other conception of the kingdom of God is pure mythology.

Christianity being the gospel of the kingdom of God, it became the religion of progress. Its aim is the growth of truth within us, and all our efforts are needed to develop truth. Thus a spiritual realm of truth and of obedience to truth, i. e. morality was created; and this spirit of progress remained the living spirit of Christianity in spite of all the vagaries of the Christian churches.

Dogmatic Christianity is dead. Yet it still exists as a dead weight. Dogmatism is barren like the thorns and thistles in the parable, and it is choking the spirit of the Christian religion, but this spirit will not die, it will spring up again and lead mankind upward and onward to higher and grander goals.

The test of progress is ever increasing truth, i. e.

an ever more comprehensive conception of the world we live in; yet the test of religion is progress.

He alone is Christ the Messiah, the saviour who leads us onward on the path of progress, and he only is a disciple of Christ who courageously follows on the path of progress. Those who attempt to make mankind stationary, who try to lock up the stream of life, and prevent the soul from growing and expanding, from increasing in the knowledge of the truth and thus developing the kingdom of God, are false prophets who come to us in sheep's clothes. They preach the letter of the gospel but suppress its spirit.

THE TEST OF PROGRESS.

The word "Progress" is one of the most commonly used terms and yet its meaning is extremely vague with most people. Progress is the ideal of our time and the glory of this generation. But what is progress? Can we give a definite and clear answer to this question, or is "progress" one of the many words by which people feel much but think little?

Progress is the act of stepping forward, it is a march onward. But who can tell us the right direction of an onward march? Did it ever happen to you when travelling on your ideal highroad of progress that you met a man who marched in the direction which you left behind? It happens very often, and if you inquire of the wanderer, Why do you go backward instead of forward? he will assure you that he marches onward while you yourself are retrogressive. Those who preach progress are by no means unanimously agreed as to the right direction. Make a chart of all the directions propounded and it will look like a compass dial. All directions possible are represented and there are not a few who believe that the development of our present civilisation proceeds in the wrong direction; they call us actually backwards to stages which lie behind us in a distant past and would consider a return to them as real progress. These retrogressive reformers are not so much among the ultra-conservative classes as among the ultra-radical enthusiasts who in one-sided idealism find perfection in the most primitive states either of absolute anarchy or absolute socialism, or whatever may be their special hobby.

The question, What is progress? is of paramount importance to ethics. For if there is no progress, if the direction of the onward movement is either indeterminable or indifferent, then, certainly there is no ethics. And if there is a special and determinable line along which alone progress has to take place, it is this alone which has to be used as a compass for our course of action. This line alone can be the norm of morality. From this alone we have to derive our moral rules, this alone can give us the real contents of the otherwise empty and meaningless term of moral goodness and this alone must constitute our basis of ethics.

Our time should know what progress is, for our generation surveys the origin and growth of life so much better than did any previous generation. We now know that all life follows certain laws of evolution and has begun from the very beginning as slimy specks of living substance developing to the present state. The man of to-day is the product of that evolution, and man's progress is nothing but a special form of evolution; it is the evolution of mankind. Our scientists have discovered the fundamental laws of evolution; so they may be able to give us a satisfactory explanation of progress. The law of evolution we are informed is adaptation to surroundings. The polar bear adapts himself in the color of his skin and in his habits to his environment; while the insects of Madeira lose their power of flight and have to a great extent become wingless. There is a survival of the fittest everywhere, but natural selection does not always favor the strongest and the best. The ablest flyers on the islands are swept by the winds into the ocean and the weak only will survive, those who are lacking in a special virtue, but not the bravest, not the strongest, not the best!

May we not imagine that there are periods or societies so radically corrupt (and history actually teaches that there were repeatedly such eras) in which the spirit of the time made it actually impossible for good men to exist and to act morally. The evil influence of tyranny, of corruption, or of hypocrisy swept the brave, the courageous, the honest, the thinking out of existence and allowed only the weak, the degenerate, the unthinking to remain? It is true that whenever a nation fell under such a blight, she was doomed. Other nations took her place and there were quite a number of peoples entirely blotted out from the face of the globe. We have progressive as well as retrogressive adaptation (as Professor Weismann informs us), and adaptation in many cases is no sign of progress in the physical world, let alone the moral progress of human beings. We may say that the law of adaptation explains survival, but it cannot afford a criterion of progress.

We will ask the philosopher what progress is. The philosopher takes a higher and more general view of life, he may give us a broader and better information as to what is the characteristic feature of progress. Progress, we are told, is "a passage from a homogeneous to a heterogeneous state." . . . "It is a continually increasing disintegration of the whole mass ac-

companied by an integration, a differentiation, and a
mutual, perpetually-increasing dependence of parts as
well as of functions, and by a tendency to equilibrium
in the functions of the parts integrated." Complexity,
it is maintained, is a sign of a higher evolution, and
it is true—in many respects higher forms of existence are richer, more elaborate, more specialised, than
lower forms. But is therefore complexity the criterion of progress; can we use it as a test wherever we
are in doubt in a special case. Does it show us the
nature of progress, its meaning and importance? It
appears that this explanation is not even generally
true, for there are most weighty and serious exceptions which overthrow the validity of this formula entirely. Is not the progress in the invention of machinery from the more complex to the less complex?
Invent a machine to do a special kind of work simpler
than those at present in use; it will, the amount and
exactitude of work being equal, on the strength of its
simplicity alone be considered superior and it will soon
replace the more complex machinery in the market.

Mr. Herbert Spencer, the philosopher of evolution, overlooked the main point when he attempted to
explain evolution as he proposed in terms of matter
and motion. Evolution means change of form, and
this change of form has a special meaning. Evolution is not a material process and not a mechanical
process, and the attempt to solve the problem of evolution on the ground of materialism or mechanicalism
(i. e. to express its law in terms of matter and motion)
must necessarily be a failure. Mr. Spencer, it is true,
recognises the importance of the formal element, for
his view of increasing complexity involves form and
change of form. Yet he selects a mere external

feature (one that is not even a universal) as characteristic of evolution and he neglects the very meaning of the change of form. This meaning remaining as an irresoluble residue in his philosophical crucible might find a place of shelter under the protecting wings of the Unknowable; but this meaning of the change of form is the very nerve of the question and all other things are matters of detail and secondary consideration.

The evolution of the solar system, being a mechanical process may find in the Kant-La Place hypothesis a purely mechanical solution. But the evolution of animal life is not a purely mechanical process. There is in it an element of feeling which is not mechanical. I do not mean to say that the nervous process which takes place while an animal feels is not mechanical. On the contrary I consider all processes which are changes of place, biological processes included, as instances of molar or molecular mechanics. But the feeling itself is no mechanical phenomenon. It is a state of awareness and in this state of awareness something is represented. This state of awareness has a meaning, which may be called its contents.

I do not hesitate to consider the meaning that feeling acquires as the characteristic feature not only of animal but especially also of intellectual life—of the life of man. It is upon the meaning-freighted feelings that soul-life originates. Let every special feeling, representing a special condition or object, be constituted by a special form of nerve-action, and we should see the soul, the psychological aspect of nerve-forms, develop together with the organism. A higher development leads naturally, as a rule but not without exceptions, to a greater complexity of nerve-forms. Yet it is not this complexity which constitutes the evolution of the soul and the progress in the development of the organism. The test of progress can be found in the meaning alone with which the feelings that live in the action of these nerve-forms, are freighted.

What is this meaning?

The different soul-forms (so we may for brevity's sake call these feelings, living in the different nerve-structures) represent special experiences and through these experiences the surroundings of the organism are depicted. The soul accordingly is an image of the world impressed into living substance and depicted in feelings. This however is not all, the soul is more than that. It is also the psychical aspect of the reaction that takes place in response to the stimuli of the surroundings. And this reaction is indeed the most important part in the life of the soul. The former may be called by a generalised name cognition or intelligence, the latter activity or ethics. The former has no other purpose than to serve as an information for the proper direction and guidance of the latter.

We do not consider the world as a chaos of material particles. We do not believe that blind chance rules supreme. On the contrary we see order everywhere and law is the regulating principle in all things and processes. The world is not a meaningless medley, but a cosmos which in its minutest parts is full of significance and purport. And this truth has found a religious expression in the God-idea. The world considered in its cosmic grandeur is divine, and when in the process of evolution the soul develops as an image of the world, the divinity of the cosmos is also mirrored in the soul. The higher animal life rises, the more does it partake of the divine, and it reaches the highest climax in men and finally in the ideal of a perfectly moral man—in the God man.

The test of progress must be sought in the growth of soul. The more perfectly, the more completely, the more truthfully the world is imaged in the soul-forms, so as to enable mankind, the individual man as well as the race, to react appropriately upon the proper occasions, to be up in doing and achieving, to act wisely, aspiringly and morally, the higher have we risen on the scale of evolution. It is not the complexity of soul-forms which creates their value, it is their correctness, their congruence with reality, their truth. Evolution sometimes leads to a greater complexity. In the realm of cognition it does so wherever discrimination is needed. But sometimes again it will lead to a greater simplicity. Complexity alone would have a bewildering aspect, it must be combined with economy, and the economy of thought is so important because it simplifies our intelligence; it enables us not only to see more of truth at once but also to recognise the laws of nature, the order of the cosmos, and its divinity.

The test of progress, in one word, is the realisation of truth extensive as well as intensive, in the soul of man. The more truth the human soul contains and the more it utilises the truth in life, the more powerful it will be and the more moral. In this way the soul partakes of the divinity of its creator, call it nature or God; it will come more and more in harmony with the cosmos, it will more and more conform to its laws, it will be the more religious, the holier, the greater, the diviner, the higher it develops and the further it progresses.

THE ETHICS OF EVOLUTION.

The first chapter of Genesis is at present interpreted by the greatest number of our theologians in a sense which is hostile to the theory of evolution. It is nevertheless one of the most remarkable documents that prove the age of the idea, for no impartial reader, either of the original or of a correct translation will find the dogma of special creation acts out of nothing justified in these verses. The first verses of Genesis tell us that God "shaped" the world beginning with simple forms of non-organised matter and rising to the higher and more complex forms of plants and animals. God shaped the heaven and the earth, is the correct translation, he made the greater and the lesser light, i. e. he formed them; he made man and the breath of man's life is God's own breath. If Darwin himself or a poet like Milton, thoroughly versed in Darwinian thought, had been called upon to present the evolution theory in a popular form to the contemporaries of Moses they could not have described it in a more striking manner. Any improvements upon the Mosaic account which could be suggested are mere trifles and matters of detail.

It is a fact that ethical aspirations, the ideal of elevating humanity, of raising men upon the higher level of a divine manhood, of creating a nobler type of human beings, of saving the souls that would go astray and showing them the narrow and strait gate which alone leads into life,—in short the *sursum* of evolution,—have been the kernel of all religions, especially those great religions which in the struggle for existence have survived up to this day—Brahmanism, Buddhism, Judaism, Christianity, and Mohammedanism. Nevertheless the idea of evolution is still looked upon with suspicion by the so-called orthodox leaders of our churches. Do they not as yet understand the religious nature of the idea? Or is it perhaps exactly its religious nature of which they are afraid? For being a religious truth, it will in time sweep away many religious errors which are fondly cherished and have grown dear to pious souls.

The idea of evolution as a vague and popular conception of the world-process is very old, but as a theory based upon exact science it is not much older than a century.

Kant told us in his "Natural History of the Starry Heavens" that an evolution is taking place in the skies, forming according to mechanical laws solar systems out of the chaotic world-dust of nebulæ. Caspar Friedrich Wolff,[*] Lamarck,[†] Treviranus,[‡] Karl von Baer,[§] and others came to the same conclusion with regard to the domain of organised life and Baer pronounced the proposition that evolution was the fundamental idea of the whole universe.[||] The work of these men is the foundation upon which Charles

---
[*] *Theoria Generationis.* 1759.
[†] *Philosophie Zoologique.* 1794.
[‡] *Biologie.* 1802.
[§] *Entwickelungs-Geschichte der Thiere.* 1898
[||] Ibid. p. 294.

Darwin stood. This great hero of scientific investigation collected with keenest discrimination and most careful circumspection the facts which prove that the struggle for life will permit only those to survive which are the fittest to live and will thus bring about not only a differentiation of species, not only an increasing adaptation to circumstances in the animal world at large, but also the progress of the human race.

The evolution in the animal kingdom has a peculiarity which distinguishes it from that of the starry heavens. It takes place exactly in the same way according to mechanical laws, being a complex process of differentiation, yet there is an additional element in it. Animals are feeling beings.

When certain motions pass through the organism of an animal there arises an awareness of the motion, and this awareness, which is a mere subjective state, is called "feeling." The same impressions produce the same forms of vibrations in the organism and the same forms of vibrations in the organism exhibit the same feelings. Every impression however leaves a trace in the system which is preserved and when properly stimulated will be reawakened together with its feeling element. When new sense-impressions are produced, the old memories of the same kind reawaken together with them, and all their feelings blend into one state of consciousness richer than the present sense-impression could be, if it stood alone and unconnected with the traces of former sense-impressions. In this way the whole world of an animal's surroundings is being mapped out in the traces left in the organism according to the law of the preservation of form, as after-effects of sense-impressions and of their correlated reactions. Many of these traces when stimulated into activity exhibit states of awareness and thus consciousness rises into existence constituting a realm of spiritual life.

This spiritual life has been called the ideal world in opposition to the world of objective reality—ideal meaning pictorial, for the ideal world depicts the real world in images woven of the glowing material of feelings.

Evolution in the animal world concentrates more and more in a development of the ideal world and this ideal world is not something foreign to the world of objective realities which it mirrors, it is intimately interconnected with it. Reality must be thought of as containing in itself the conditions of bringing forth feeling beings and through feeling beings the ideal world; and this ideal world is not merely a phantasmagoria, a beautiful mirage without any practical purpose, it is to the beings which develop it the most important and indispensable thing, for it serves them as a guide through life and as a basis for regulating their actions. If the world of objective realities is correctly depicted in the ideal world, it will help them to act in the right way, so as to preserve their lives, their existence, their souls. Ideas which are correct, which faithfully represent the realities which they depict, are called true, and actions which are based on and regulated by true ideas are called right or moral.

Thus the ideal world contains in germ the possibilities of truth and of morality.

Evolution in the spiritual world means the development of truth, it means an expanse of the soul, a growth of the mind as well as a strengthening of the character to live in obedience to truth.

When Mr. Spencer undertook to write a philosophy of evolution, he was fully conscious of the sweeping importance of the evolution theory, but when he approached the ethical problem, he became inconsistent with his own principle and instead of establishing an ethics of evolution, he propounded an ethics of hedonism regarding that action as right which produced the greatest surplus of pleasurable feelings.

Pleasurable feelings are experienced under most contradictory conditions. Pleasures cannot form any standard of ethics or a regulative principle to guide our appetites. Pleasures on the contrary are often dangerously misleading and many a life has been wrecked by trying to choose that course of action which promises a surplus of pleasures.

Feelings are mere subjective states and their importance depends entirely upon the meanings which they convey. It is not the pleasurableness of feelings and of ideas which ought to be considered when they are proposed as norms for action, but their correctness, their truth. That which brings man nearer the truth and harmonises our actions with the truth is right, and that which alienates man from the truth is wrong. Accordingly that which makes our souls grow and evolve is moral, that which dwarfs our souls and prevents their evolution is immoral.

There is but one ethics and that ethics is the ethics of evolution.

FAIRY TALES AND THEIR IMPORTANCE.

The attempt has been made to banish fairy tales from our nurseries. The cry is raised "away with ogres and fairies, away with fictitious monsters! Let us teach our children truth and nothing but truth. Prepare their minds for life. It is a downright injury to fill their imagination with stories that are unreal, untrue, and even impossible."

This proposition is made on the ground that everything unreal is untrue; therefore it is obnoxious and should not be allowed to be instilled into the minds of children.

The principle of removing everything untrue from our plan of education is unquestionably good. The purpose of education is to make children fit for life, and one indispensable condition is to teach them truth, wherever we are in possession of truth; and, what is more, to teach them the method how to arrive at truth, how to criticise propositions, wherever we have not as yet arrived at a clear and indisputable statement of truth.

Allowing that fairy tales are unreal and may lead the imagination of children astray: are they for this very reason untrue? Do they not contain truths of great importance, which it is very difficult to teach children otherwise than in the poetic shape of fairy tales? I believe this is the reason why in spite of so much theoretical antagonism to fairy tales they have practically never been, and perhaps never will be, removed from our nurseries. There are no witches who threaten to abuse the innocence of children, and there are no fairies to protect them. But are there not impersonal influences abroad that act as if they were witches, and are there not also some almost unaccountable conditions in the nature of things that we meet often in the course of events, but which act as if they were good fairies to protect children (and no less the adult children of nature called men,) in dangers which surround them everywhere, and of which they are not always conscious?

Science will at a maturer age explain such mysteries, it will reveal to the insight of a savant that which is a marvelous miracle to the childish conception of an immature observation. But so long as our boys and girls are not born as savants, they have to pass through the period of childhood, they have to develop by degrees and have to assimilate the facts of life, they have to acquire truth in the way we did, when we were children, as the race did, when humanity was in a state of helpless childhood still.

Did not religion also come to us in the form of a fairy tale? And is not a great truth contained in the legend of Christianity? The belief in the fairy tale will pass away, but the truth will remain.

The development of children, it has been observed, is a short repetition of the development of the race. Will it be advisable to suppress that stage in which the taste for fairy tales is natural? Is not a knowledge of legends, fairy tales, and sagas an indispensable part of our education, which, if lacking, will make it impossible to understand the most common place allusions in popular authors? Our art galleries will become a book with seven seals to him who knows nothing about the labors of Hercules or the Gods of Olympus. Will you compensate the want of an acquaintance with our most well-known legends, sagas, and characters of fiction at a later period, when the taste for such things has passed away?

I met once an otherwise well-educated lady who did not know who Samson was. An allusion to Samson's locks had no meaning to her, for she had enjoyed a liberal education, her parents being freethinkers, she had never read the Bible and knew only that the Bible was an old-fashioned work, chiefly of old Hebrew literature, which she supposed was full of contradictions and without any real value.

A total abolition of fairy tales is not only inadvisable, but will be found to be an impossibility. There are certain classical fairy tales, sagas, and legends, which have contributed to the ethical, religious, and even scientific formation of the human mind. Thus not only many stories in the Old and the New Testament, but also Homer, Hesiod, and many German and Arabian fairy tales have become an integral part of our present civilization. We cannot do away with them without at the same time obliterating the development of most important ideas. Such fairy tales teach us the natural growth of certain moral truths in the human mind. These moral truths were comprehended first symbolically and evolved by and by into a state of rational clearness.

I do *not* propose to tell children lies, to tell them stories about fairies and ogres and to make them believe these stories. Children, having an average intelligence, will never believe the stories, however much they may enjoy them. The very question: Is that really true? repeated perhaps by every child, betrays their critical mind. Any one who would answer, "Of course, every word is literally true," would be guilty of implanting an untruth in the young minds of our children. We must not suppress but rather develop the natural tendency of criticism.

While we cannot advise the doing away with fairy tales, we can very well suggest that the substance of them may be critically revised, that superfluous matter may be removed and those features only retained that are inspiring and instructive.

THE VALUE OF MYSTICISM.

MYSTICISM is the blight of science. Mysticism in science is like a fog in clear daylight. It makes the steps of the wanderer unsafe and robs him of the use of his most valuable sense—the sense of sight. There is impenetrable darkness around him; everything is confused by insolvable problems. The whole world appears to the benighted mystic as one great and inscrutable enigma.

Mysticism in religion is widely different. It is here where the value of mysticism must be sought for. But religious mysticism does not claim that truth is unknowable. It claims not only, as does science, that truth can be *known*, it claims that truth can be *felt* even before it is known. Truth is a strong and wholesome power, unconquerable and omnipotent, which is available not only to the knowing but to those also who grope in the dark, yet cherish the love of truth in their hearts.

A scientist can scientifically enquire into the social laws, and can after a life-time of long and laborious study arrive at the truth, that what is injurious to the swarm is not good for the bee. The ethical maxims: thou shalt not steal, thou shalt not kill, thou shalt honor father and mother, the scientist will perceive, are not cunningly invented by religious or political leaders, they are the indispensable conditions under which alone society can exist. Wherever they are not heeded the whole community will go to the wall. The individual that sins against these laws will injure society, yet he will ruin himself at the same time.

The ethical truths are important truths, and it is good to know them, to understand their full importance. Yet even those who are unable to grasp them in their minds; those who have not the scientific knowledge to see how the moral law works destruction to the trespasser and is a blessing to him who keeps the law—even the unscientific, the poor in spirit, can feel the truth; they can trustingly accept it on faith and *can be sure* that they are right. And truly, if they do accept it, if they act accordingly, they are better off than those scientists who have arrived at some approximations that upon the whole it is perhaps after all even for the single individual better to be honest, than to be shrewd.

There are scientists and among them some of great name and fame, who after a life-time of long and laborious study did not arrive at the ethical truths that the moral commands will preserve, and that they do preserve, both the individual who keeps them and the society to which that individual belongs. There are naturalists who are very familiar with a certain province of nature, especially with the brute creation. They say, not the morally good will survive, but the strongest, the cunningest and the shrewdest. The naturalists who say that, are most learned professors; they are crammed with biological data, and have made many zoölogical observations; they know facts of nature and have classified them as natural laws—but Nature herself has not revealed her divine face to them. They have not entered the holy of holies in the temple of Creation, for they see parts only, and do not perceive the whole; they overlook the quietly working tendencies of the whole. They misinterpret the meaning of the partial truths that happened to come under their observation.

Moral truth can be felt. Therefore let religious mysticism gain hold of man so as to make him feel the truth of the moral law even before he is able to understand it.

The moral feeling is man's conscience. The moral law and man's trust in the truth of the moral law must not be planted into the reasoning faculty of man only, it must be planted by example and instruction into his heart long before the reasoning faculty of his mind is developed. It must be made part of his inmost soul long before he commences to study, to learn, and to observe. It must be the basis of his whole being, and the determining factor of his will.

If the moral law were merely superadded in later life, if its presence in our minds rested upon abstract conclusions only, upon logical arguments and syllogisms, how uncertain, how precarious would its influence be upon our lives. Rational insight must come to strengthen the moral truth of our soul, but its roots must be deeply buried in the core of our heart. Science will come to explain what conscience is, and why conscience is right in this or in that case, science will also assist us to correct an erring conscience, but if the basis of a man's character has not been laid in early childhood, science will come too late to benefit him through moralizing arguments. A conscience that is grounded upon ratiocination only, is weak in comparison to a conscience that permeates the whole being of a man, his emotions, his will, and his understanding; his heart as well as his head. Conscience must be, as we say in popular speech, our "second nature"—yea, it must be our "first nature," so that in all situations of life, in tribulations, and in temptations it will well up unconsciously with an original and irresistible power, even before we can reason about the proper course of our actions.

The tempter approaches us always in the name of science, but his arguments are not science, they are pseudo science. The tempter says: "Do not be foolish, be wise. The criminals are convicted not for their crimes but because they were fools; they were not shrewd enough to escape the consequences of their deed. Be wise, be cunning enough, and thou wilt outwit all the world." There is no criminal who did not think himself wise enough to escape the law, and if he regrets at all, he will commonly regret not the deed but one or the other of his mistakes which, as he supposes, betrayed him. The criminal tries to remove the vestiges of his deed; yet the acts done to this purpose become new and powerful witnesses against him. They, chiefly, become the traitors that deliver him to the judge.

Do not be deceived by the pseudo-wisdom of your thoughts that lead you into temptation. They will lead you into ruin, if you follow them. Do not be deceived by the escape of evil-doers from their legal punishment; they carry a punishment within them which is worse than the penitentiary. Neither be deceived by the success of the unprincipled. Many of those whom you suppose to be morally depraved, are perhaps after all not so unscrupulous as you think. They may have virtues and abilities, strength of will, power of concentration, industry, intelligence, foresight in business combinations, of which you think little, but which meet the wants of their time and serve the common good. Such men succeeded, perhaps, in spite of those faults in their characters to which you erroneously attributed their success. If they are really unprincipled, and are successful in their enterprises, do not judge of them before you have seen the fulfillment of their destiny.

The royal psalmist of Israel, the shepherd boy, who was a poet and at the same time a hero, who became the king of his nation because he treated even his enemies with justice, had during his career often seen the unprincipled succeed, and so he sang:

I have seen the wicked in great power and spreading himself like a green bay tree.

But David continues:

Yet he passed away, and, lo, he was not; yea I sought him but he could not be found.

Mark the perfect man and behold the upright, for the end of that man is peace.

It may seem to you as if crooked means were better than straightforward truth, as if small trickery and well-calculated deceptions would gain the victory over the simplicity of honest dealing. It may seem so to you and it may seem so to your friends and advisers. It is not! Truth and justice are always stronger than the strongest lies. And if you do not understand it, believe it and act accordingly.

I do not mean to say that if your cause is just, if you are morally good and honest in your purpose, that truth and justice will come down like gods from heaven to assist you. O, no! You must fight for truth and you must stand up for justice with all your abilities and foresight. What I mean to inculcate is not blind confidence in the victory of truth and justice, as if they intended actually to appear on earth to work for you, instead of your working for them : I mean to say that, under all circumstances, falsity, untruth, injustice, and all immoral means, however cunningly they may be devised, are the most dangerous allies. Whoever associates with them will be sure to go to wreck and ruin. The way to success, to a final and solid success is only that steep and thorny path on which virtue led the Greek hero to Olympus. Because strait is the gate, and narrow is the way which leadeth unto life and few there be that find it.

THE UNITY OF TRUTH.

Truth, thou art but one. Thou mayest appear to us now stern and now mild, yet thou remainest always the same. Thou blessest him that loves thee, thou revealest thy nature to those that seek thee, thou hidest thy countenance from him that disregards thee, and thou punishest him that hateth thee. But whether it is life or death thou givest, whether thy dispensations are curses or blessings, thou remainest always the same, thou art never in contradiction with thyself; thy curses affirm thy blessings, and thy rewards show the justice of thy punishments. Thou art one from eternity to eternity; and there is no second truth beside thee.

There was a strange superstition among the learned of the middle ages. The Schoolmen believed in the duality of truth. Something might be true, they maintained, in philosophy, which was not true in theology; a religious truth might be true so far as religion was concerned, but it might be wrong in the province of science, and *vice versa* a scientific truth might be an error in the province of religion.

The *Nation* of August 7th, 1890, contains a criticism by an able pen of the aim which is pursued by *The Open Court*. But the criticism is written from the standpoint that the duality of truth is a matter of course; whereas it is merely a modernised reminiscence of the scholastic doctrine that that which is true in science will not be true in religion.

We are told:

"The profession of *The Open Court* is to make an 'effort to conciliate religion with science.' Is this wise? Is it not an endeavor to reach a foredetermined conclusion? . . . Does not such a struggle imply a defect of intellectual integrity and tend to undermine the whole moral health? . . . Religion, to be true to itself, should demand the unconditional surrender of free-thinking. Science, true to itself, cannot listen to such a demand for an instant. . . . Why should not religion and science seek each a self-development in its own interest?"

It is true enough that many religious doctrines stand in flat contradiction to certain propositions that have been firmly established by science; and the churches that proclaim and teach these doctrines do not even think of changing them. There are dogmas that defy all rules of sound logic, and yet they are retained; they are cherished as if they were sacred truth. But church doctrines and dogmas are not religion; church doctrines and dogmas are traditions. They may contain many good things but they may also contain errors, and it is our holy and religious duty to examine them, to winnow them so as to separate the good wheat from the useless chaff.

Let us obey the rule of the apostle, to hold fast only that and all that which is good. And what is good? Let us inquire of Truth for an answer. That is good which agrees with truth. Good is not that which pleases your fancy, however lofty and noble your imagination, and however better, grander, or sweeter than the stern facts of reality you may deem it to be. You will find that in the end all things that appear good, but are not in accord with truth, are elusive: they will be discovered to be bad; usually they are worse than those things which are bad and appear so to us at first sight.

What is religion? Religion is our inmost self; it is the sum total of all our knowledge applied to conduct. It is the highest ideal of our aspirations, in obedience to which we undertake to build our lives. Religion in one word is truth itself. Religion is different from science in so far as it is more than scientific truth; it is applied truth. Religion does not consist of dogmas, nor does the Religion of Science consist of scientific formulas. Scientific formulas, if not applied to a moral purpose, are dead letters to religion, for religion is not a formulation of truth, but it is living the truth. True religion is, and all religion ought to be what Christ said of himself and of his mission, "the way, the truth, and the life."

If a teacher tells his pupil never to be satisfied with his work until the result when examined agrees with the requirements, and to work his examples over until they come out right; is that a predetermined conclusion? In a certain sense it is, but not in the sense our critic proposes. If objection is made to a duality of truth, and if it is maintained that religion and scientific truth cannot contradict each other, is that an effort which "implies a defect of intellectual integrity and tends to undermine the whole moral health"? Just the contrary; it is the sole basis of intellectual integrity, it is the indispensable condition of all moral health.

"Religion to be true to itself should demand," and that religion which *The Open Court* proposes, does demand not "an unconditional surrender of free-thinking" or of free enquiry, but an unconditional devotion to truth. Does science demand free-thinking?

Perhaps the answer may be "yes," and there can be no objection provided that free-thinking means free enquiry and the absence of all compulsion. But the free-thinking that is demanded by science means at the same time an absolute obedience to the laws of thought. The same free-thinking, which is at the same time an unconditional surrender to truth, is the cardinal demand of religion. The great reformer Martin Luther called it the freedom of conscience and considered it as the most precious prerogative of a Christian.

The Open Court does not propose to conciliate science with certain Christian or Mosaic or Buddhistic doctrines. This would be absurd and such an undertaking would justly deserve a severe criticism, for it would be truly a predetermined conclusion in the sense that our critic intends. It would "imply a defect of intellectual integrity and undermine the moral health." Autocracy and individualism are not reconcilable, but socialism and individualism are reconcilable. Order and liberty are not such deadly enemies as may appear at first sight. Superstition and science are irreconcilable, but religion and science are not irreconcilable. Indeed, the history of religious progress is a constant conciliation between science and religion.

Religion and science, it is maintained, must "seek each a self-development in its own interest." Certainly it must, but this does not prevent that which we deem to be religious truth being constantly examined before the tribunal of science, and that which we deem to be scientific truth being constantly referred to religion. Our critic seems to have no objection to religion and science coming into accord, but he proposes to wait until they approach completion. If this maxim were universally adopted, there would be no progress in the development of religion. Is not "completion" a very relative state? Waiting for completion would be about equivalent to stopping all social reform until mankind has reached the millennium. Every social reform is a step onward along the path to the millennium, and every conciliation between science and religion is a step onward in the revelation of living truth.

The religion of the middle ages was a religion of dualism, it proposed the duality of truth. The religion of the future will be a religion of Monism; and what means Monism? Monism means unity of truth. Truth is invincible. It never contradicts itself, for there is but one truth and that one truth is eternal.

LIVING THE TRUTH.

They are but few who do the thinking of mankind, and the great masses are led by the few sometimes in the right, sometimes in the wrong direction. It matters little whether this is to be regretted or not, it remains a fact and must be faced. Yet this state of things makes every independent thinker the more valuable. Every man who is an independent thinker is a power in his sphere, and will contribute a share to the further evolution of thought in humanity.

The intellectual battles of mankind are mostly fought out by a few leaders, and the great mass is ready to follow those who have been successful in the fight. Nevertheless we must recognise that thought has increased; and there are many unmistakable symptoms that humanity is making progress at an increasing ratio. This lets us hope that the misery unnecessarily and foolishly produced by improvidence or ignorance will be lessened and that knowledge will spread together with a general good-will among men. This is the aim of thought, nay it is its necessary result.

Thought is not mere sport. Thought is the most important, the most practical, the most indispensable activity of man. Thought is the savior of mankind,. and the salvation of man is the goal of the aspirations of those who struggle against superstition and indifference.

I do not hesitate to say that indifference is worse than superstition. I am always glad to meet a thinking man who is earnest in his defense of some old creed, if he is only honest. However much I may differ from his views I shall always treat him with the respect due to sincerity. Difference of opinion must never induce us to set aside justice; and after all a man who is sincere and has an independent conviction, even though his conviction be utterly wrong, does a greater service to progress than the indifferent man who will always belong to that party which happens to be the fashion of the day. Indifference more than error hinders progress.

I see the thinkers of mankind, few though they are, divided into two camps. The champions of the one trust in progress and work for constant amelioration; the champions of the other believe that innovations are extremely dangerous, and the best thing for humanity would be to remain stationary. Those of the latter class will concede perhaps that in the domain of industry and in the sciences progress must be made, but they do not believe in the progress of religion. Their religion is to them perfection, it represents in their minds absolute truth, and progress of absolute truth, progress of something that is already perfection, is, as a matter of course, gilding refined gold.

The battle waxes hot between the two parties, the former is strong through its alliance with scientific aspirations, but the latter is still in the majority. It is in possession of the great mass of indifferent people; and the champions of progress may often become despondent so as to give up all hope of a final victory. Ignorance seems stronger than knowledge and folly more powerful than wisdom. In a moment of such despair Schiller is said to have exclaimed: "Against stupidity fight even Gods in vain."

Who among us when confronted with unconquerable superstitions, has not had such sentiments at one moment of his life or another? And now I ask, can we know which party in the end will be victorious? Can we know the means by which alone a victory is to be achieved? Let me in a few words indicate the answer which I trust is very simple in the general plan of its main idea, and yet so very complex in its application that we could philosophize on the subject as long as we live. Indeed, mankind does philosophize on the subject and has never as yet got tired of it. And I suppose it never will, for here lies the object of all science, of all knowledge, of all philosophy.

What will conquer in the end? Truth will conquer in the end. By what means will truth conquer? By being truth, or in other words by morality. That party will conquer, be it ever so weak in numbers, be it ever so badly represented, that is one with truth. But it is not sufficient merely to know the truth. Truth must be lived.

Only by living the truth shall we be able to conquer the world. Therefore it is necessary to recognize the all-importance of morality. The ethical problem (as I have often said on other occasions) is the burning question of the day. To know the truth, to preach the truth, and also to denounce the untruth of superstitions is very important; but it is more important to live the truth.

If you have two men, one of whom knows the truth but does not live it, while the other lives the truth but does not know it; who must be regarded as nearer the truth? Surely he who ignorant of the truth lives it, and not he who knowing the truth does not.

What is truth? Truth is agreement with the facts of reality. Truth accordingly is not a mere negation of untruth, not a mere rejection of superstitions. Truth is positive, it is the correct recognition of facts as well as of the laws that live in the facts and have been abstracted therefrom by science. Morality is the agreement of our actions with truth, and the most important truths for the regulation of men's actions are the laws which rule the relations between man and man forming the conditions of human society.

The strength of the many organisations that still hold to antiquated superstitions lies in the fact that after all they try their best to obey the moral laws. And the weakness of many free-thinking persons as well as organisations, lies in their neglect of ethics, They do not feel the urgency of demanding strictness in morals; they are perhaps not exactly immoral but they are indifferent about the claims of morality.

The moral laws have been formulated by Religion first in mythological expressions; but the mythology of Religion is slowly changing into a scientific conception of facts. Mythology is fiction, it preaches the truth in parables. Nevertheless it contains actual truth. And the religious parables are not less true, they are more true than the unthinking believers imagine. The truth of these parables is grander, subimer, higher than the similes in which they are expressed.

Here lies the secret of success. The church has grown into existence and has attained its power because it was the ethical teacher of mankind in the past. On the one hand it appears that the church refuses to progress, and on the other hand progressive thought has heretofore too much neglected to become practical or in other words to push the moral applications of truth.

We stand now before a crisis: Either the churches will reform; they will cease to believe in superstitions; they will acknowledge truth and the correctness of the scientific methods of reaching truth; in one word they will become secular institutions, institutions adapted to the moral wants of the world we live in; in which case they will remain the ethical teachers of mankind; or those institutions which represent progressive thought and have recognized truth and the rational means of reaching truth, will more and more inculcate the practical applications of truth; and if they do, *they* will become the moral teachers of mankind.

Truth must conquer in the end; but knowing the truth is not as yet sufficient; it is living the truth which will gain the victory.

THANKSGIVING-DAY.

As the sun rises to-day from the depths of the Atlantic, he beholds a great and prosperous nation celebrating one of the most beautiful festivals of the year. It is the day of giving thanks for all the bounties which Nature, our common mother, has showered upon us in the year gone by. It is the day of giving thanks for the rich harvest now being gathered into the barns of the farmer, and which we who are not farmers, shall none the less enjoy. For all of us, the merchant and the artisan, the manufacturer and the banker, the artist and the scholar, the soldier and the sailor, all of us who make an honest living, depend ultimately on the blessings that Nature bestows upon us, the fruits that grow in the fields, and the meat that she provides.

It is true that we must work for it. In the sweat of our face we must eat our bread. But all our labor would be in vain if Nature ceased to yield the harvest which in abundance she annually offers.

* * *

Considering the state of affairs in this light, we must have a feeling of pride and at the same time of modesty. Of pride, because our prosperity, our property, our life with all its future hopes, are the result of our own work ; what we are is the product of our own and our forefathers' endeavors. Of modesty, because all our labor would be in vain if that omnipotent power of natural forces did not continually carry along upon its mighty billows of life the courageous boats of thinking beings.

We must learn to know, that what we are, we are through nature only; for we ourselves are but parts of that great power in which we live and move and have our being.

Our fathers in their gratitude called that power of omnipotent Nature God, and Christ taught us to revere it in child-like love as a Father. If we have ceased to believe in a humanized Deity, if we no longer adopt the idea of a personal God, we must not forget that there is a great truth in the words of the psalmist who sings:

> Except the Lord build the house, they labor in vain that build it; except the Lord keep the city, the watchman waketh but in vain.
>
> It is vain for you to rise up early, to sit up late, to eat the bread of sorrows: for so he giveth his beloved sleep.

It is a noble feature in man's nature that prompts him to celebrate great events and to remember the momentous days of his existence. But our feasting must not consist of good eating and drinking alone. Our festivals must be a consecration of our life. Festivals, if celebrated in a truly humane spirit, will elevate man's actions by thought and ennoble his work by reflection.

> " 'Tis that alone which makes mankind—
> And 'tis the purpose of man's reason
> That he consider in his mind
> His handiwork of every season."

You who are happy, you who look back upon a year that has yielded its harvest, rejoice in the blessings of

Nature, rejoice in the health of life, rejoice that you behold this day! Be thankful for the bounties you have received and close not the doors of charity to the needy and the poor who are less fortunate than yourselves!

The unfortunate, the sick, the poor are invited to join in the general joy and to rejoice in the general prosperity of our country, in the glorious growth of our nation, and in the noticeable progress of all mankind which apparently leads more and more to higher and purer ideals of the universal brotherhood of man.

Those who are prosperous will celebrate this sacred day with a grateful mind, sympathetic towards those who are stricken with the many ills that flesh is heir to. Let us remember our own weakness, let us consider that what we are we are not of ourselves. Thus we shall learn the wisdom of modesty that teaches us to look upon the forlorn and shipwrecked as brothers, so that we shall lend them a helping hand. Let us assist the fallen and downtrodden in the right spirit, not in the arrogance of our own merits, of our own good luck and fortune, but in the fraternal love of a pure-minded and heartfelt kindness.

* * *

Blessed be the sun that shines upon this day, and blessed be his return in all future years. Blessed be the country that yields us the fruit upon which we live, and blessed be that great nation that flourishes in this wonderful land of liberty. May the highest ideals we cherish, be realized in her destinies!

CHRISTMAS.

The Christmas bells will soon chime and with their harmonious peals they will bring joy and merriment into every household. There is a secret charm in the celebration of this holy festival. It is wonderful what sacred gladness attaches to the sight of the glorious tree that remains green in winter-time, when it is decked with glittering ornaments and its many candles shed their joyous light upon the circles of frolicking children with roseate cheeks and beaming eyes!

What is the mystery of this jubilant feast, and how is it possible that wherever it has been introduced, there it will remain as the dearest and most cherished of all holidays?

First Christmas was celebrated as Yule-tide by the old Teutons, especially by the most northern tribes of the great Teutonic family, the Norsemen and the Saxons, as the return of the sun, as salvation in midst of anxieties and troubles, as the victory of light over darkness. As many other feasts so Christmas, and Christmas, it seems, more than others, is a festival of natural religion. Then the Christians adopted it and very appropriately selected it as the memorial day of the birth of the Saviour. Now it is celebrated by Christians and Pagans, by Jews and Gentiles, by all who came in contact with Saxons or Germans, or their kindred in the North. No one can withdraw from the sacred spell that the worship of Nature exercises even now upon our minds. Christians like to forget that their Christmas tree is an old pagan symbol of the world. It is Ygdrasil, under the branches of which the three norns of the present, the past, and the future are sitting, lisping runes and weaving the fates of the Universe. There is Urd's well at the roots of the holy tree and its water is sacred. The norns spray the water upon the branches of Ygdrasil which sinks down into our valleys as dew. This keeps the tree ever green and strong.

The festive Yule tide has been a holy season to our Teutonic ancestors since times immemorial; since they settled in their northern homes in Europe, which their descendants, the Norwegians, the Danes, the Dutch, the English, and the Germans still inhabit. The dreariest days of the year, when darkness and frost with snow and ice were most oppressive, became by reaction as it were the most joyful time.

In the northern parts of Norway the sun disappears entirely towards the close of December, and when after an absence of two nights or more it rose for a short time on the horizon, it was saluted with bonfires lit with yule-logs, with festive processions, with fir-trees illuminated with candles, with merry-making and family feasts of all kinds.

The mistletoe which grows on holy oak-trees and remains green in winter-time, whose seed was supposed to have fallen from heaven, was the sun-god Baldur's sacred plant. With mistletoe therefore the houses were decorated, and the greeting under the mistletoe was all love and friendship in the name of Odin's fairest and most righteous son. Baldur had been killed by the dark and gloomy Hœdur, but he was restored to life again. With Baldur all nature received new life, and all mankind rejoiced in him.

When Christianity was introduced, how could a better day for the celebration of Christ's nativity be selected than Baldur's festive day. The birthday of Jesus was not celebrated in the early church, and there is not even the faintest legendary account regarding its date. Our Teutonic ancestors succeeded in settling this problem in favor of their dear Yule-tide by a quotation from the scriptures. John the Baptist says as to his relation to Christ: "He must increase but I must decrease." (John iii. 30.) Accordingly, St. John's day was fixed upon the 24th of June when the days begin to decrease, and Christ's upon 25th of December when the days begin to increase again.

Yule-tide lost none of its charms when it was changed into Christmas. On the contrary, the sacred joys *Weihnacht* gained in spiritual depth and importance, preserving all the while the old pagan ceremonies that symbolize the immortality of light and life.

Christmas is not a feast of any special creed or nationality. The custom of celebrating it has spread from the Teutonic nations to France, and Spain, and Italy, and Ireland, and over the whole world. It is now the family feast of almost all mankind whether they believe in Jesus as their saviour or not.

We keep the Christmas season as a dear and sacred time which in the midst of a dreary winter night reminds us of the sun's return. Darkness cannot conquer light, and death cannot conquer life. Christmas teaches us to bear up bravely in troubles, to keep up hope in misfortunes, to preserve the courage of life in the midst of struggles of cares and worries, and to spread joy around us so far as it is in our power.

There are times so dreary that in our anxiety we see no hope but death. There are days so bleak and wintery that we begin to despair, and encumbered with cares we cry, "The evil is stronger than the good in this world, and the power of darkness quenches the glory of light." The days become shorter and shorter. The nights become longer and longer. A general corruption is prevailing and increasing; the moral sense is growing debased and retrogression seems all but universal.

O ye of little faith! Be of good cheer, and in the midst of all your trouble and worry celebrate a joyous Christmas. For Christmas is the commemoration of the holy morn that greets us after the longest night. It reminds us of the undying hope, that light and life are eternal. It is true that life is a world of woe, full of toil and of pain. Nevertheless, there is a saviour born into the world; and this saviour is the son of man. The ideal son of man lies as yet in the cradle. But we know that he will grow; he will rescue the world from those troubles which are caused by folly and crime; he will elevate mankind through purity and justice; and he will consecrate life and the struggle for life through the noble aims which more and more will become conscious ideals in the minds of men.

REVELATION.

In my childhood I was told that there were two kinds of divine revelation. God had revealed himself (1) in Nature, and (2) in the Scriptures. Neither revelation was easy to decipher and interpret, but God always aids the endeavors of the upright, and the one revelation would assist us in understanding the other.

There is, too, according to the catechisms, a third kind of revelation : the Conscience of Man. Man has an instinctive recognition of that which is right and that which is wrong, and this instinct is sometimes a most wonderful and accurate guide, although there are many cases in which it leads astray. Conscience, we are told, is the voice of God, and the behests of conscience we are bound to obey, although we must be on our guard lest conscience be perverted by errors and superstitions.

These three revelations of God must be one and the same. If they are true and reliable they must agree, and wherever they do not agree our interpretation of one of them, or of two, or of all them, is wrong. As a matter of fact, we find that the three conflict, and we must accordingly investigate which of the three is the most reliable.

The dogmatic Christian claims that the Bible is the most reliable; and in all religious matters the

Bible must be considered as the ultimate authority. Yet, whatever precious doctrines the Bible may contain, it can be considered as divine only in so far as it is true, and God cannot proclaim one truth in nature, and another truth in the Scriptures. He cannot be one God to all the world, and another God to a few prophets. God might reveal himself more fully to those who are maturer in mind, whose souls are further advanced in moral and mental growth, for God reveals himself to the extent that we search for him, and are able to comprehend the truth. Yet the two revelations should never be contradictory. They might be different in degree, but not in kind.

Of the three divine revelations there is but one that is consistent, one that never contradicts itself, that has remained unchanged, and will remain so forever. That is the revelation of God in Nature. There is order in nature, and law rules supreme. All natural phenomena are in all their glorious variety so many instances of the oneness that pervades nature, and among all the natural phenomena, the most wonderful revelation of God appears in man ; and in that which is most human in man, in language, and in thought. Every truth is divine, every truth is a revelation, and every scripture thus inspired will prove useful in working out righteousness. Therefore we agree with the apostle when he says :

> Every scripture inspired of God is also profitable for teaching, for reproof, for correction, for instruction, which is in righteousness: that the man of God may be complete, furnished completely unto every good work.—II. THIM., 3, 16. 17.

It is not the Bible alone which is a revelation of God, but the Vedas, the Zendavesta, Homer, the Koran, the Edda ; Shakespeare, and Goethe ; and

Kant and Darwin, and all the scientists. All the scriptures, all the literatures of all people so far as they contain thoughts that are noble and elevating, and beautiful and true—they are all revelations of God. In so far as a book contains errors it is not devine, it is no revelation of God, whether it be incorporated in the biblical canon or not.

The Bible was considered by the old Hebrews in this light, for the Old Testament is nothing but a collection of the Hebrew literature up to a certain date. Had Goethe lived among the Jews at the time of David, and had the anachronism been possible that he had written his Faust at that time; Goethe's Faust would be one of the canonical books in the Bible of to-day.

Conscience, it is true, is a revelation of God; but what is conscience but the development of the ethical instinct in man.

Experience has taught man that certain acts that promise to be pleasant at first, will cause regret afterwards; that the injury done to others will not bring to him the benefit he expected, but may even entail harm which he never thought of. Experience will teach him that self-denial and unflinching love of truth, even where they appear very obnoxious, will in the end prove to be the best. Conscience accordingly is ultimately based upon experience, not only of ourselves, but of parents and teachers. It is partly an inherited tendency; partly it is based upon all the remembrances of our life from earliest childhood. The examples given us by beloved and respected persons, by our elders and by our friends, are written in our souls and will consciously and unconsciously influence our actions. It is neither uncommon nor strange that the voice of man's conscience is often perverted, by bad examples and insufficient or wrong instruction. As the knowledge of the medicine man is the rude beginning of science; so is conscience a natural product which needs refinement and culture by methodical education.

The only direct and reliable revelation of God is to be found in the facts of nature; and all the other revelations in the Scriptures, and in conscience, are but parts of this one and only true revelation. They are true only in so far as they agree and represent this; and the truth of this can be revised again and again. The book of nature is open to every one, and in the places where to-day we understand its disclosures imperfectly, we can hope that to-morrow by more careful observations and closer investigations, we shall better comprehend its meaning.

Truth is the exactness with which the harmony of cosmic order is represented in the mind of a thinking being; truth is the mark of divine dignity in man, through truth and truthfulness we become children of God, and truth is the saviour of all evil.

GOD.

Who is God and what is God? is a question that is raised by both religious and irreligious people; and most different answers are given. Every one of us has perhaps his own and peculiar opinion about God; some of us are theists, some pantheists, some atheists, and there are in the history of religion and philosophy, so far as I can judge, not two thinkers who fully agree upon the subject. Shades of differences are visible everywhere.

I do not intend to discuss any one of the many conceptions of God; nor do I intend to preach either Theism, or Atheism, or Pantheism. All I ask is the use of the word God in the sense of "the ultimate authority in conformity to which man regulates his actions." Of those who allow their actions to be determined by the first impulse that comes over them, I would say, that whim is their God. Those who are swayed by egotism, we say that self is their deity. There are others whose sole principle of conduct is the pursuit of pleasures: their God is happiness. Others still may possess a moral ideal; the endeavor to be obedient to their duties: their God would be duty.

After this preliminary definition of God, we put the question: Is there any way of ascertaining the nature of God, so that all men of different opinions may be led to the recognition of one God, who is the only true God, beside whom all other Gods are mere idols? In other words, Is the authority in conformity with which man regulates his conduct merely his private pleasure, is it purely subjective in its nature, or is it a power that is above us, that is mightier than ourselves, that enforces obedience and wrecks those who dare to disregard it? Is that saying of Antisthenes true, "The Gods of the people are many, but the God of nature is one?"

The answer to this question is simple, and can easily be deduced from experience. I cannot at all act as I please, but have to regulate my actions according to the facts of nature. If I attempt to walk on the water I shall sink; if I try to fly from the top of my house to the roof of my neighbor's house across the street, I shall fall. Natural laws will not be altered on my account, and I shall not be able to fashion them so as to suit my purposes. However, I can accommodate myself to the facts of nature, I can obey the natural laws, and if I do so, it will be to my own benefit. The more intimately man is acquainted with nature, the more perfectly he adapts himself to the order of nature, the wider will be his dominion. In the measure in which he becomes more obedient to the authority of natural laws, the more powerful, the more independent, the more free will man be.

Schiller said:

> "Within your will let deity reside
> And God descendeth from his throne."
>
> "[Nehmt die Gottheit auf in euren Willen
> Und sie steigt von ihrem Weltenthron.]"

The natural laws of the physical world, gravitation, mechanical laws, physical laws, biological laws, may appear to the present generation plain and palpable facts of nature, yet it took centuries to sum up the facts in laws and to state some of them in simple terms. The men who succeeded in stating them in simple terms were prophetic geniuses, such as Copernicus, Kepler, Galileo, Newton, Huyghens, Darwin, and others; the results of their labors are discoveries of a divine inspiration, and are a revelation of the eternal and universal order of nature.

Besides the physical laws of nature, there are the sociological laws that prevail in the higher kingdoms of living organisms, and in the societies which greater numbers of single individuals unite. Every one of us is a member of a community; and again all the communities of human beings are closely bound together, however great the distance in which they dwell, by certain relations, by common interests, and mutual sympathies. These sociological laws are not a product of well calculated intentions, but they are of a natural growth; the evolution of the social affairs of mankind is deeply rooted in the conditions of things.

Now every fact of science stated as a law has its practical side; it teaches us how to behave in certain conditions. There is no knowledge but it can be framed in the shape of a moral command. The tables of arithmetic are mere statements of fact; but every one of them is a most valuable ethical law: it is a guide for our actions and a rule of conduct.

Every child knows that the ethics of arithmetic cannot be changed, it is a sovereign power above us. Yet we can make that royal authority descend from its throne by obedience to its behests, we can adapt our calculations to it, and thus we shall partake of its sovereignty. The more accurately and the purer truth dwells in our minds, the more will our souls grow divine, and the more will we bear in ourselves the image of God. There is no knowledge that does not make us purer, and no correct application of knowledge that does not make us more divine. But among all the natural laws that it behooves a man to know and to obey, are the laws of human life, the relations among human beings, and the aspirations of human ideals. It is here where the revelation of God appears in its grandest, its most beautiful, and its holiest form.

How many people are there that understand that these laws are no less cogent and irrefragable than the laws of the multiplication tables! How many imagine that they can break these laws with impunity. Let us do evil, they say, that good may come from it.

The prophet Hosea says: "People are destroyed from lack of knowledge," and these words are true even to-day. People injure themselves and others mostly from ignorance and from ill-will, which is a necessary result of ignorance. Would not the brute cease to be brutish if it were endowed with human reason?

Let us open our eyes to see and prepare our minds to learn the ordinances of the divine authority that shapes the destinies of our life. The better we observe them, the clearer we understand them, and the more promptly we obey them, the sweeter will be the blessings that come upon our lives, the greater will be the advance of humanity, and the nobler will appear the divinity of mankind.

DESIGN IN NATURE.

At a meeting of a scientific club lately, a discussion was held on the subject: "Is evolution directed by intelligence?" This question touches the very heart of religion and science; and we cannot shirk it if we desire to attain to any clearness and comprehensiveness of view concerning the most vital problems of human existence.

Before we can answer the question proposed, we must first ask what do we understand by intelligence. We must analyze its meaning and separate it into the elements of which it consists.

Intelligence comprises two elements: (1) We mean by intelligence design, plan, order, harmony, conformity to law, or *Gesetzmässigkeit;* and (2) when speaking of intelligence we think that there is attached to it the element of feeling or consciousness.

Feeling by itself has nothing to do with intelligence; yet consciousness has: consciousness is intelligent feeling. A single feeling, a pain or a pleasure, as long as it remains isolated cannot be called intelligent; yet it acquires a meaning as soon as it refers to one or several other feelings. For thus feelings become representations of the surrounding conditions that produce feelings. Consciousness is nothing but a co-ordination of many feelings into one harmonious state. Beings in possession of conscious intelligence we call persons.

Now we ask, Can there be design which is not connected with feeling? Can there be order or plan without a conscious being who made the plan? We say, Yes.

The crystallization of a snowflake is made with wonderful exactness, in agreement with mathematical law. Is this formation of snow-crystal manufactured with purposive will, by a personal being? A mathematician knows that the regularity of forms necessarily depends upon the laws of form, upon the same intrinsic order which is present in the multiplication table; it depends upon the arithmetical relations among the numbers.

Is a personal intelligence necessary for creating the laws that produce the harmony of arithmetical proportions? Is a personal intelligence necessary for making the angles of equilateral triangles equal? Certainly it is not.

Suppose that some substance crystallizes at a given angle. Necessarily it will form regular figures shaped according to some special plan.

Suppose again that certain cells of organized substance, plant-cells or animal-cells, perform special functions, will they not in their growth exhibit a certain plan in conformity to their nature not otherwise than a crystal? They will, or rather they must; or can we believe that the interference of personal intelligence is necessary to apply the plan to the growth of organized substance? Organization is so to say crystallization of living substance; it is growth in conformity to law.

The growth of a child takes place unconsciously,

not otherwise than the growth of a flower. The consciousness developed in the former is the product, not the condition of its development; it is the product of organization. The consciousness of man is the highest kind of systematic co-ordination of feeling that we know of, and therefore we say that he is endowed with intelligence. Man is a person.

Personality is not the annihilation of the mechanical law; yet through the introduction of feeling the mechanical law that governs the changes and innumerable adaptations of a person, becomes so complex that it at first sight appears to us as an annihilation of the mechanical law.

The hypothesis of a personal intelligence is not needed to explain either the design of nature, or the plan of evolution, or the gradual development of nations and individuals, which processes are all in rigid conformity to law. At the bottom of all cosmic order lies the order of mathematics, the law that twice two is always four.

Personal interference is so little necessary to produce regularity according to some design with any exactness, that it would even make it all but impossible. If man desires the execution of some work with minute exactness, he has to invent a machine to do the work. A machine performs its work with rigid immutability. And a machine, what is it but an unfeeling and an unconscious,—a mechanical,—intelligence? Personality, what is it but the power of constantly renewed adaptation? Personality, therefore means mutability.

Suppose a book were written and not printed; suppose it were produced by the conscious intelligence of a personal being, and not mechanically by a machine;

could we expect the same minute exactness? Assuredly not. It would be witchery to adapt anything in close and rigid conformity to law, without machine-like unconscious intelligence.

Suppose that the planets were run by some personal being; that they were constantly watched with conscious wisdom and regulated by purposive adjustment; we could not trust our safety a moment on this planet. Mechanical regularity in minutest details is all but impossible in the work of personal intelligence.

* * *

A machine has no feeling and possesses no conscious intelligence; yet a machine must have been invented by a conscious and premeditating intelligence. A machine proves the presence of a designing person somewhere. And the question arises: Could not the Cosmos be considered as a machine invented by a great and divine person, designed for some preconceived end?

Even though there were no objections to this rather child-like and antiquated anthropomorphism, this conception of things would be of no use towards explaining the cosmic order. A machine is not invented by an inventor as a fairy-tale is conceived by a poet. A machine can work only if it conforms to that impersonal intelligence which we call mathematical necessity. It is the latter that makes the machine useful, and it is the latter that has to be explained.

If God made the world as an inventor makes a machine, he had to obey the laws of nature and to adapt his creations to the formulas of mathematics. In that case, however, the Creator would not be the omnipotent and supreme God; there would still be an impersonal Deity above him. In that case the Creator would be no less subject to the cosmic order than we poor mortals are.

Show me by any convincing argument that the cosmic order represented in so simple a statement as "twice two is four" had to be created arbitrarily by some conscious intelligence, and I shall willingly and without hesitation return to the anthropomorphic belief in a personal God—a belief which was so dear to me in my early youth. Yet so long as the cosmic order must be recognized as uncreated and uncreatable, as omnipresent and eternal, as omnipotent and irrefragable, we must consider the worship of a personal God as pure idolatry.

* * *

But this solution of the problem—is it not dreary atheism? It is not, or it is—according to our ability to receive the message of the necessity, the irrefragability of the Formal Law.

Our theologians maintain that the order of the cosmos proves the existence of a deity. I maintain that it does more: The order of the Cosmos is itself divine. It does not prove that there is a God outside the universe who made the cosmic order; it proves the presence of a God inside.

Is the order of the Cosmos void of intelligence? It is without feeling, but surely not without plan or design. The laws of nature represent design; they are embodied design. The law of gravitation, for instance, does not act with consciousness, yet it represents order. It describes the regularity of the fall of a stone as well as of all the motions of the heavenly bodies in their wonderful order.

The immutability of the cosmic order disproves a supernatural God, but it proves an immanent God.

And this God cannot be a person. He is more than a person. God is called in the Old Testament the Eternal, he is represented as immutable. Can a person be immutable? Is not personality embodied mutability, is it not adaptability to circumstances? The divine order of the Cosmos as represented in Natural Laws stands above all mutability—unchangeable, inadaptable, eternal.

* * *

This God, the immutability of impersonal, or rather of superpersonal intelligence, is the condition of science and the basis of ethics. If natural laws were personal inventions which could be changed at the pleasure of their inventor, science would become impossible, and morality would become an illusion. What is morality but our effort to conform to the order of nature, and above all, to the laws that shape society?

This impersonal intelligence is higher than personal intelligence, as much so as the laws of a country are infinitely higher and holier than all its citizens, its princes and kings not excepted. There is a rule in monarchies that the sovereign stands above the law. Is it necessary to explain that this idea is a farce, an illusion, a felony against the sanctity of the law? Similarly, the idea of a God, fashioned according to the personality of man, is a blasphemy of the higher God, of that God who alone is God, of the Deity that passeth all understanding, *i. e.*, all conscious reasoning and personal wisdom.

The worship of a personal God is the last remnant of paganism. Our religious convictions can and will not be purified until we apperceive a glimpse of the grandeur of a higher view.

There is a superhuman Deity, whose glory the

heavens declare, and the firmament showeth his handiwork. Day unto day uttereth speech, and night unto night showeth knowledge. There is no speech nor language where their voice is not heard. The whole Cosmos is permeated by eternal and divine law, by intelligence, by design.

The whole world is a glorious revelation of its immanent God. Yet this revelation is concentrated in man's personality. He possesses, not only a conscious intelligence reflecting in his soul the divinity of the All, but also the aspiration of moral ideals inspiring him to conform to the cosmic order that rules supreme from Eternity to Eternity.

THE CONCEPTIONS OF GOD.

AMONG the conceptions of God there are three which have been and are still the most prevalent and powerful; these three are Theism, Pantheism, and Atheism.

The Theist anthropomorphises that power which he recognises as the authority of moral conduct, and looks upon it as a stern ruler or a kind father. If evils appear as the consequence of vice, he says: These are God's visitations! And he thinks of God as teaching his creatures his will and enforcing his obedience, not by making the contrary absolutely impossible, but like a wise educator raising children in liberty, allowing them to make mistakes so as to learn by their own experience.

Theism is not wrong if we keep before us the fact that the personality of God is an allegory; and it must be granted that it is the best allegory we can discover. There is a world-order manifesting itself to those who have eyes to see and ears to hear. We have to conform to it and there is no escape from it. It is omnipresent, like all natural laws; like gravitation it is everywhere, it is bound up in all existence, being that something that encompasseth all our life.

In describing this omnipresence of God, the psalmist says :

Whither shall I go from thy spirit ? or whither shall I flee from thy presence ?

If I ascend up into heaven, thou art there : if I make my bed in hell, behold thou art there.

If I take the wings of the morning, and dwell in the uttermost parts of the sea,

Even there shall thy hand lead me, and thy right hand shall hold me.

There has been made, so long as Christianity exists and even longer, a strong opposition to the idea that God is, like man, an individual being, having at different times different passions and desires. The Old Testament contains the well-known passage : "God is not a man that he should lie ; neither the son of man that he should repent."

God is as little a person as are the ideas of Goodness, Beauty, and Truth ; and the passages of the Bible in which God is described as wroth or repenting, or as being subject to any emotion or sentiment of a human character, have been understood since they were written, by rabbis no less than by the fathers of the Church, in an allegorical sense, which was not only appropriate because of the strength and expressiveness of the simile, but because it was also the language of the time. To speak or think of spiritual things otherwise than in the habits of the times would be equivalent to expecting that the author of Genesis should have known Darwin's origin of the species and all the details of natural history when he described in great poetical outlines the formation of the world and the origin of man out of the dust of the earth.

The dogmatic view that God is a person and must be considered as a person became finally established as the orthodox view of the Church during the second and third century after Christ, and in this way all other views were branded as atheism. But who gave to a few narrow-minded bishops and to the theologians of a special school the right to impose this interpretation of the Bible upon all mankind? Who gave the right to Athanasius to pronounce as an œcumenical confession of faith the *Quicunque vult salvus esse*, i. e. "No one can be saved except he believe as is here prescribed." Living the truth can save alone. But the truth cannot be pronounced on the motion of a bishop by the majority decision of an ecclesiastic council. The truth must be searched for, it must be established by careful observation and critique, it must be proved.

We are willing to recognise the truth wherever we find it, even in the errors of the past; we will patiently winnow all opinions and creeds, lest we throw away the wheat together with the useless chaff. But with all that, we do not intend to compromise with superstitions sanctified by traditions. If Athanasius's view of God and other religious conceptions are to be regarded as infallible truth too sacred for criticism and required to be accepted blindly, we shall openly and squarely side with atheism and denounce the belief in God as a superstition.

Atheism is right in the face of dogma and dogmatic theism. There is no person ruling the world; all the processes of nature take place with an intrinsic necessity according to the life that is in everything that exists. The whole world is one great cosmos pervaded by unalterable law.

But was the idea of God not something more than a belief in a huge person? Is it possible that an enormous error swayed the intellectual development of humanity for millenniums? The strength of the God idea was not its error but its truth, and its truth is contained in the fact, that in spite of the advantages which sin, malevolence, iniquity, falsehood, and disregard of the rights of others seem to bring the evil-doer, humanity still believed in the final victory of justice and the triumph of truth. And this one feature in the idea of God was predominant whenever and wherever it exercised a moral influence over the minds of men. It gave them strength in temptation, hope in affliction, and confidence in tribulation. And shall we relinquish this treasure because it was alloyed with error? Shall we drop with the personality of God all the moral truth which the idea contains?

Schiller says:

> "One God exists, one holy will,
> While fickle man may waver.
> Above time and space there liveth still
> The highest idea forever."

If, then, God is no person, if God is considered as the All in All, if Nature alone is God, is not the latter view nearer the truth than theism? This view which identifies God and the world is called Pantheism, and it cannot be denied that in the face of the theistic view, pantheism is a deeper and more correct conception of God. Nevertheless, Pantheism has also its blind side, and most of its defenders are entangled in gross errors.

It is true that the idea of a personal God outside of the world and nature is not tenable; yet the idea of God and the idea of nature are not identical. God is nature in so far only as nature serves us as a regulative principle for our actions. God is the cosmos in so far only as its laws represent the ultimate authority of moral conduct. God is not the heat of the sun, not the rain that descends from the clouds; he is not the blossom of the tree, nor the ear of wheat in the field. The idea of God is a special abstraction, different from other abstractions, and it should not be confounded with them. Pantheism recognising the truth that there is no God outside of the universe, preposterously confounds God and the universe and thus leads to the confusion of a God-Nature, in which there is no wrong, no sin, no evil.

It has been said, and it is true, that the weakness of Pantheism is its inability to explain the evil of the world. If the All is in every respect absolutely identical with God, there is no evil: if everything is a part of God, its existence whatever it be, even the existence of evil, is sanctified by being divine. There would be no wrong, but there would be no right either. The morally bad would disappear together with that which is morally good, and the whole would appear as an absolutely indifferent and meaningless play of physical forces.

Does this state of things really represent life as it is? Are there no ideals, no aspirations? Is there no direction, no goal, no aim in the evolution of life and in the development of mankind? Surely there is good and bad, there is right and wrong, there is health and sickness, there is prosperity and ruin, evolution and dissolution, building up and breaking down; there is heaven and hell in human hearts, there is God—and the devil. The world as it is is possible only in these contraries, in these oppositions, and its life is a constant struggle between Ormuzd and Ahriman.

It is a vain dream to think of a world which is good throughout. We can as little think of light that casts no shadow as of "good" without being the resistance to "evil," or without standing in a contrast to "bad."

Christ said :

"Woe unto the world because of offences! For it must needs be that offences come ; but woe to that man by whom the offence cometh."

The Talmud contains a legend that the rabbis had once succeeded in catching the devil and keeping him confined, when lo! the whole world came to a standstill. Everybody went to sleep and all life ceased. Suppose it were possible that a world existed without any evil, it would be a world without any opposites, it would be a world of indifferent homogeneity, without aim, without direction, without interests. If there were at all in an absolutely good world a play of forces evolution would be as good as dissolution, progress would be equivalent to retrogression, and the cosmos would be a machine which might be turned backward just as well as forward.

Could you have a thermometer which indicates the heat only and not the cold at the same time? Good and evil are relations which are deeply founded in the nature of things. These relations arise through the very complications of life. To identify God and the All, to understand by God the upward direction just as much as the downward direction of evolution, is the same mistake as to identify the concepts heat and temperature. It is true that the same degree of the thermometer may now be perceived as heat and now as cold. Heat and cold are not two things mixed in our temperature ; they are one. So are good and evil. Nevertheless there is a difference in the rising and the falling of the thermometer. There is a difference of heat and cold. This difference is relative and it disappears as soon as we leave the sphere of relations and consider either a single moment in its unrelated isolation or the total whole in its absolute entirety. A single act in my life if it remained unrelated and isolated could be called neither good nor evil. There is no absolute evil; nor is there any absolute cold. An isolated act would be like a certain position of the thermometer of which we do not know whether it represents a rise or a fall. It becomes hot or cold not until it is referred to another state of temperature. And there is no sense either in speaking of the morality or immorality of the All in its absolute totality.

That which appears to us from our standpoint as evil—and I do not deny that, considered in this relation, it is actually and undeniably evil—appears if considered in the whole as a part of the total development of universal life, as a transitional and a necessary phase only. It is a partial breakdown, but it is no absolute destruction.

The evil in the world is comparable to the negative magnitudes and quantities in arithmetic. There are no negative things in the world; but there are negative magnitudes in arithmetic. They represent a contrary direction to that which has been posited. The minus is a positive operation, but this operation is employed to reverse a plus of equal magnitude. The plus and minus operations have sense and meaning only if considered in their mutual relation. This relation being neglected we have only single operations or the results of operations, but neither positive nor negative magnitudes. If the impossibility could be thought, that there are no interconnections among the parts of the whole cosmos, we should have neither bad nor good, but only isolated actual existences.

Consider the whole world as a whole and destruction disappears as much as new creations. There are, so far as we can see, only actual existences which move onward somehow in some direction. That which appears to us as a dissolution, as a destruction, is in the motion of the whole a mere preparation for a new generation. The breakdown of a solar system must appear only as an evil, as a negative operation in comparison to the positive operation of a building up. But in the entire cosmic life it will most likely be the indispensable preliminary phase of the construction of a new world. In the entire cosmic life, there is no evil, there is the progress of formation on the one hand and there is on the other hand the dissolution of those combinations which have become unfit for a continued existence. They must be dissolved in order to be prepared for new formations; and thus their dissolution may be considered as a blessing, as much as the curses that rest upon sin, if viewed as integral parts of the whole world-order, are not inflictions; they are as much blessings as the gains that accompany noble deeds.

In this sense we may say that God is everywhere in nature, he is in evolution, he is in dissolution, he will be found in the storm; he will be found in the calm. He lives in the bliss of good aspirations and in the visitations that follow evil actions. He lives in the growth of life and in its decay. God is not the storm, he is not the calm, he is not the decay of life, he is not dissolution. He is not the bliss of virtue, nor is he the curse of sin. But he is in them all.

In contradistinction to Theism, Atheism, and especially to Pantheism, we call this conception of God Entheism.

God is the indestructible *sursum*, which ensouls

everything that exists, which constitutes the direction of evolution and the growth of life, which is the truth in the empire of spiritual existence. It is an actuality, no less than matter and energy; and indeed like these two, which represent as it were God's reality as well as his power and omnipotence, it cannot be lost in all the changes that take place in the constant formation, dissolution, and re-formation of solar systems. It is eternal, and it is in him we live and and move and have our being.

IS GOD A MIND?

We read in the first chapter of Genesis:

"And God said, Let us make man in our image, after our likeness: and let them have dominion over the fish of the sea, and over the fowl of the air, and over the cattle, and over all the earth, and over every creeping thing that creepeth upon the earth.

"So God created man in his own image, in the image of God created he him."

These verses are significant. They have a scientific meaning. To us God is that power to which we have to conform; he has produced man such as he is, that is as the thinking being that aspires to ever higher and nobler ideals, to us accordingly the view that man is created in the image of God becomes self-evident and almost tautological. But primitive thinkers starting from the supposition that man is a likeness of God were led to the strange error that God in his turn must be a likeness of man. Thus arose all the anthropomorphic conceptions of God.

That power which produced man—let us at present call it "nature" so as to avoid the old confusion of anthropomorphism—cannot have been matter and nothing but matter, it cannot have been force or energy and nothing but force, it cannot have been sentiency or the conditions of sentiency, and nothing but potential sentiency. Nor can it have been form or a formative principle alone. It cannot have been law and order only. It must have been all this together. Matter, force, sentiency, form, law, and order are only aspects of nature, they are only abstract ideas representing some qualities of reality, which alone is the One and All. And this One and All is not a meaningless chaos, as it represents itself in minds that are confounded, but an orderly and living whole bringing forth out of itself sentient beings in whom its existence is mirrored. Existence mirrored in minds is not a mere *Fata Morgana*, a beautiful mirage, but it serves the practical purpose of guidance, to let the children of nature live in accord with its great mother, to show them the way of salvation, the gate that leadeth unto life.

When we speak of nature we think as a rule of certain single phenomena only of this One and All; we think of mountains and trees but not so much of man's mind and his interferences with the rest of nature—for properly considered man's mind is a part of nature. When we speak of reality, we think above all of its actuality, its efficacy, its immediate presence, but when we speak of God, we think of it as an authoritative existence, as our standard of ethics, as the moral law, allegorically represented as our Father, that is, as the power that created us and guides us still, to which we have to conform in our ethical aspirations. Nature, Reality, God, or whatever other expression we may have for the One and All of the great Cosmos in its infinite manifestations and in its eternal being, are all names only, abstract ideas representing now this and now that quality of one and the same existence.

Sentient creatures, the children of God, in so far as they are psychical are called minds. And we ask, What do we understand by minds?

A mind, in brief, is a description of the world in ideas. "Ideas" means literally "images." The different things are represented, and the interaction among these representations is called thinking.

How ideas originate is a question the solution of which can only be hinted at in this connection. Mind can originate only in feeling beings. The feelings of feeling beings are different according to the different sense-impressions through and with which they make their appearance, similar sense-impressions being associated with similar feelings. Thus feelings acquire meaning. The various causes of the different sense-impressions are symbolised in various feelings as well as in the memory pictures of these various feelings. Ideas again are symbols representing whole groups of such feelings as are somehow constantly associated. And the glorious evolution of the realm of ideas in living beings is easily explained if we consider its usefulness as a means of information concerning the surrounding world. They afford the possibility of orientation and serve as a guidance for action. With the assistance of representative images plans of action become possible, and a conception of a better arrangement of this or that state of things—generally called an ideal—is of the highest importance to the further development of life and mind. A growth of mind leads to an increase of power. Each acquirement of truth means an expanse of the dominion of mind in nature.

Minds naturally grow by degrees; they start with simple feelings in irritable substance, and in the long run of millenniums through a preservation of soul-structures (generally called hereditary transmission) and, in the higher grades of life, through a direct transference of mind by means of education they gather a rich store of soul-structures, of pictures representing innumerable objects as well as the subtle relations among these objects.

Let us now ask whether God can be a mind. Our answer is decidedly negative. Every mind is a world of representations, of pictures, of ideas; and these ideas, pictures, and representations have a meaning. If they are true they represent realities. Now if there is a God, and we say that there is, God is not ideality but reality; he is not a mental representation of the actual world, of nature, of the Universe, of the Cosmos; he is much more than a mere representation, he is the actual world, nature, the Universe, the Cosmos itself. He is the One and All, not a part of it, or a mere picture of it. God is also the picture, and he is that quality of the world which makes the picturing in minds possible. God is in the mind, he reveals himself in the human soul; he appears in Truth. But God is not only the truth; he is infinitely more than the truth, he is the reality represented in the truth.

Truth is truth because it is an image shaped unto the likeness of the original. The human mind is created as an image of God. Now the theologian comes and says, Man is like God, man is mind—i. e., a world of images or ideas—therefore God must be a mind. Is this not like saying, This is a picture of George Washington, it is like George Washington. Therefore George Washington is a picture! No! George Washington is more than a picture; he is the original of the picture!

It is often said that man is a finite mind and God is an infinite mind. But what has either infinitude or finiteness to do with mind? Mind, every mind, is infinite in its possibilities, there is no limit to its growth, there is no boundary which it cannot reach and transcend. But at any special state, as at present or at any moment in the future, mind is and always will be something definite. Consider that all mental representations are possible only through limitation. Thus vision is possible only through focusing the eyes upon one spot. Comprehension in mental pictures, is a focusing of the mind's attention upon one thing or one feature of things. Accordingly minds in this sense are always finite, always limited. Every mind is always the mind of a concrete being and the contents of every mind are also of a concrete kind. Think of infinite pictures, or infinite ideas! What a meaningless combination of words! If God, the One and All, is infinite indeed, he certainly cannot be a mind.

We might and some people indeed do understand by mind the nature of mind, mentality. The nature of mind may be found in sentiency or in that quality of nature which produces sentiency— we call it potential sentiency. Or it may be found in the order prevailing among the mental representations, which order is representative of the objective world-order, of the cosmic law and the rationality of the universe as represented in cosmic laws. Very well. If "mind" means the nature of mind, then certainly God is mind, but he is not *a* mind.

If God were a mind, it were necessary for him to have ideas. Otherwise his mind would represent without representations and symbolise without symbols. He would have to think his ideas consecutively as we do and form different associations at a time. Yet, what would mental representations avail him? He need not think, he need not speak to himself in order to make up his mind to act in this or that way. He simply acts. He in his all-sufficiency is always himself and thus he is consistent with himself.

In the catechism this truth is mythologically expressed in the idea of omniscience. Nature, as it were, obeys the law everywhere. The falling stone falls as if it knew the law of gravitation and had correctly computed the present case. Nature need not know the law in order to obey it. She need not employ the symbols of mental representation to remain consistent with herself. She is herself everywhere, and the laws of nature are a part and feature of nature. We say, Nature is as it were omniscient. Actually nature is more than omniscient. As omniscient, she might communicate information about all things of herself to herself. This communication, however, is so direct, she being herself everywhere, that its means, i. e. the symbols, which are the crutches of communication, disappear into zero. The communication is received before it is pronounced.

That God should be the One and All, and at the same time a mind, would be something like saying, that a man in order to be a man and himself, should always have his passport or his picture in his pocket. No! If we speak of the man, we mean the man and not his picture. If we speak of God, we mean the All-Being and not a mind, we mean the original and not the copy, we mean the creator and not the creature.

Is it Atheism to deny that God is a mind? If you understand by God that he is a person like ourselves, it certainly is Atheism. But if the conception of God as a mind and a person were the only allowable God-idea, then theism would be paganism. What is paganism but the personification of parts of nature or nature as a whole and the acting accordingly. Pagans try to bend the course of nature and natural laws not by their own efforts and honest work, but by prayers and sacrifices—as if God or the Gods were human beings like ourselves influenced by flatteries and bribable by gifts! Christ has done away with the vain repetitions as do the heathens, but the Christians still cling to Pagan customs, pagan rites and a pagan conception of God.

People who have given little thought to the subject might think, that if God is not a mind, it is as good as if he did not exist. Then he would only be brute force and crude matter. But this is a mistaken conception of God. The materialist runs to the other extreme. God is not mere force and God is not crude matter. How grand and divine this wonderful All-Being is, can only be learned from its manifestations. The heavens declare the glory of God and the firmament showeth his handiwork. Day unto day uttereth speech and night unto night showeth knowledge. There is no speech nor language where their voice is not heard. Yet grander than all the starry heavens in their glorious concert is the soul of man, the mind that yearns for truth, the spirit that understands, and aspires to achieve, the work of truth.

The All, the Cosmos, God, or by whatever name we may call the great whole of which we are parts and phenomena, is not a heap of material atoms nor a chaos of blind forces. The most characteristic feature of his being is order and law. And this order and law is called in the New Testament Logos—i. e. rationality, reason, logical consistency. God would be no God without the logos. This Logos is a constitutional part of God. God is not a mind, but he is mind, he is logos, and he appears in mind. God is not truth, but he appears in truth. This is the revelation which Christianity has brought into the world.

Says St. John: "In the beginning, [that means from eternity] was the Logos and the Logos was with God and the Logos was God. All things were made by him and without him was not anything made that was made. . . . And the Logos was made flesh."

This last sentence is the kernel of Christianity. The divinity of the world appears in humanity, and and true humanity embodies all that which we call divine. The son of man is the child of God and the ideal of humanity is the God man. God is not *a* mind, but nevertheless God is mind, and when we come to ask, where is the Father, Christ answers very positively and unmistakably "I and the Father are one."

Those who believe in God, as being *a* mind are more pagan than they are aware of. It may be said that God is mind, but not *a* mind. Suppose he were *a* mind, is that not actually polytheism only with the number of Gods reduced to the singular? Christ does not say, God is *a* spirit, but "God is spirit." Yet the pagan conception of God has been so influential that the translator has inserted that little word which changes a most radical, a philosophical and a monistic idea into the very same superstitions against which Christ had protested so vigorously.

Science is not dangerous to religion, and clear thought is not against the teachings of Christ. Science is dangerous to superstitions and clear thought is incompatible with many dogmas and conceptions which are upheld at present by the Christian churches. The dogmatist rightly shuns the light of science, but the religious man, that is, he who wants truth unadulterated and is ready to conform to truth, to live it and to act according to his best knowledge of truth, he will not lose his religion but purify it through thought and scientific exactness of thought.

Says Lord Bacon:

> "A little philosophy inclineth Man's mind to atheism, but depth in philosophy bringeth men's minds about to religion."

Bacon's view of God is not clear and thus this famous saying of his also lacks lucidity. We understand it and quote it in the sense, that a little philosophy is sufficient to make apparent the contradictions and absurdities contained in the traditional idea of God. But a deeper insight will reveal the profound truth that is contained therein. Depth in philosophy will help us to purify the fundamental conceptions of religious thought, above all the idea of God. When we maintain that God is not a mind, we do not deny that he is mind, taking mind in the sense of the Greek "logos"; and at any rate he is greater than the greatest human or other mind can be, for he is the reality itself of which a mind is only an image, a symbol, and a representation.

IS THE INFINITE A RELIGIOUS IDEA?

Prof. Max Mueller's view of religion is based on the conception of the infinite. His idea of God is the infinite behind the finite. He says:

"Convince the human understanding that there can be acts without agents, that there can be a limit without something beyond, that there can be a finite without a non-finite, and you have proved that there is no God."

Is this not going rather too far? Does the agent supposed to be behind the processes of nature constitute nature's divinity? Prof. Max Müller's view of God is scientific as well as radical, but it makes of religion a metaphysical speculation; it identifies it with the conception of an hypothetic something behind nature of which we really know nothing. It appears very desirable to free religion from this metaphysical element and build it upon the positive facts of our experience which will always remain its safest foundation.

Positivism knows of no agent behind the natural phenomena; it dispenses also with the agent behind the psychical processes of soul-life. Positivism is an economy of thought. Instead of viewing acts as motions produced by the pressure of an agent behind them, we think the act and agent together as one. The agent is *in*, not *behind* the act. The act is the agent itself.

Positivism is commonly represented as atheism just as much as the view of the orthodox Oxford Professor would have been decried as atheism some ten or twenty years ago. And I grant that Positivism is not Theism, if Theism means the belief in a personal God who being shaped into the image of man, is conceived as an individual being, as a great world-ego swayed by considerations and even by passions and emotions, thinking now of this now of that thought, and regulating the affairs of the universe as it pleases him like a powerful monarch.

There is nothing more or less divine in the infinite than in any other mathematical, logical, or scientific idea. The infinite has one advantage only—if it be an advantage—over other ideas; its nature is less understood. But if there were anything divine in the conception of the infinite, why do we not use such formulas a $\frac{0}{1}$ or tangent 90 degrees, or simply the sign ∞ as holy emblems in our churches?

Prof. Max Müller must have felt this insufficiency of the idea of infinitude as the basis of religión. At least he has on another occasion modified his definition. In another article of his,* Prof. Max Müller says:

"It may be said in fact it has been said, that the definition of religion which I laid down is too narrow and too arbitrary. . . . I thought it right to modify my first definition of religion as 'the perception of the Infinite,' by narrowing that perception to 'such manifestations as are able to influence the moral character of man. I do not deny that in the beginning the perception of the Infinite had often very little to do with moral ideas, and I am quite aware that many religions enjoin what is either not moral or even immoral. But though there are perceptions of the Infinite uncon-

---

* *Fire-Worship and Mythology in their Relation to Religion*, (*The Open Court*, page 2322, No. 146, Vol. IV.—16).

nected as yet with moral ideas, I doubt whether they should be called religious till they assume a moral influence. On this point there may be difference of opinion, but every one may claim the right of his own opinion."

The infinite, it appears to me, is not at all a specially religious idea, and it will be very difficult to prove how the idea of the infinite can ever assume a moral influence, except in a very limited sphere. The powers of nature in their overwhelming influence upon the fate of man in a beneficent and evil way, the light of the sun, the flashes of the thunderstorm, the joy of great triumphs, the enthusiasm after extraordinary successes, our trials and sorrow at the bedside of our beloved ones, the agonies and anxieties of life, in one word definite and actual realities have done much more than the idea of the infinite in the production of religion. I am aware that Prof. Max Müller says: "These finite realities suggest an infinite agent beyond them." But this is no description of religion; it is an interpretation of religious ideas, representing them in a special phase of development.

The infinite may have produced a religious awe in a lonely scholar when he pondered over the problems of its nature and found himself unable to solve them. And it may have stirred a still deeper religious emotion in the mathematical mind who succeeded in solving some of its problems. But the same religious influence must be attributed to any other scientific idea. Was not Kepler overwhelmed with the grandeur of the cosmos when he solved the riddle of the motions of the heavenly bodies? Was not his emotion truly religious, and is there anything infinite in his formulas?

It will be noticeable that the infinite as a properly

religious idea enjoys a very limited field. The two greatest religious documents are to my mind the Decalogue representing the Old Testament and the Lord's Prayer representing the New Testament; in neither can any idea of the infinite be found. It is true that the Lord's prayer ends with the clause "for thine is the kingdom, and the power, and the glory, forever, Amen." "Forever" I grant, means infinite time. But it is well known that these words are not genuine with Christ; they have been added by the Christians of the first or second century; and if they were genuine, how incidental is the idea of the infinite, how secondary if compared with the momentous propositions of the prayer itself! It appears that religion would not suffer if the idea of the infinite were entirely dropped from its definition and Prof. Max. Müller's additional clause (i. e. "that which will influence the moral character of man") were made its main essence.

The definition of God as the infinite conveys no clear idea. The popular view of the infinite is very indefinite, and its scientific conception is a thought-symbol for a process never to be finished. The scientific view of the infinite does not represent a complete and real thing, but an incomplete and never to be completed function. Suppose that in measuring the world we arrived at the last star of the farthest milky way and took our stand between the definite reality behind, and empty space before us, is there no divinity in the finite existences we have measured, and is God living in the nothingness of the infinite space that lies beyond us unmeasured and immeasurable?

Let us define God as those realities of our experience to which we have to comform; as those manifestations of nature which we cannot fashion; as those laws of cosmic existence which we have to obey; and atheism will never again rise to overthrow the proofs of an existence of God. God is the authority of moral conduct, and religion is the basis of morality. All ideas which influence the moral character of man are religious, while dogmas are either religiously indifferent, as if they represent ideas having no bearing upon moral conduct, or even deeply irreligious, if they are productive of immoral habits. And one of the most immoral church doctrines, not as yet entirely abandoned by orthodox people, is that man should believe blindly. It is a sacred religious duty to investigate the truth most scrupulously. Religion is not belief in the supernatural as the theologian of the old school says, nor is it the search for the infinite, as Prof. Max Müller says. Religion is much simpler. It is our search for truth with the aspiration to regulate our conduct in accord with truth.*

* Prof. F. Max Müller wrote to the author with reference to the above criticism of the Infinite as a religious idea: "I thank you for your article on my fourth Lecture. I quite agree with your objections, and when you see the whole of the lectures, you will find how carefully I guarded against this misapprehension. The Infinite is simply the highest generalisation for all that ever formed the object of religion. There is no wider term, it is wider even than Spencer's Unknowable, as I tried to show. But here as elsewhere we want a katharsis of language, otherwise we shall never have a new philosophy. F. MAX MUELLER."

GOD, FREEDOM, AND IMMORTALITY.

KANT showed in his Critique of Pure Reason that the ideas Soul, World, and God are 'paralogisms of pure reason.' We can arrive at these concepts by a logical fallacy only. We may nevertheless, he declared in his Critique of Practical Reason, retain these concepts, because they are of greatest importance for our practical and our moral life. If we act as if we had no soul, and as if no God existed, we are more likely to go astray than if we act as if we had an immortal soul and as if a God existed—a God, a just and omnipotent judge, who will reward the good and punish the evil.

Upon the need of morality he builds an ideal world, the foundations of which are the ideas of *Freedom* (including moral responsibility), *Immortality*, and *God*. Being fully conscious of the fact, that these ideas are not provable, Kant called them "the three postulates of practical reason."

The conflict between Pure Reason and Practical Reason proves that in Kant's philosophy traces of Dualism are preserved which lead him to incompatible assertions. He boldly and honestly lays down the inconsistency of his philosophy in his four "antinomies," or contradictory statements. Popularly expressed, they are:

THESIS.	ANTITHESIS.
1. The world is limited.	1. The world is infinite.
2. The soul is a simple substance, and therefore immortal.	2. The soul is a compound, and therefore destructible.
3. There is moral freedom distinct from the law of causality.	3. There is no freedom, but all is subject to causality.
4. There is a God.	4. There is no God.

Kant believes that the arguments to either issue, the positive or the negative, are of equal weight. Thesis as well as Antithesis, he declares, can be defended or attacked with equal force.

Is it not strange that a great man can fall into so great an error—an error that is at the same time so palpable? Of two statements that are contradictory, one only can be true. It is impossible that both are right, or that the arguments of either are correct. Yet it is possible that both are wrong, that the formulation of the dilemma is radically incorrect,—and is such the case with Kant's antinomies.

We resolve the four antinomies into the following statements, which cannot be said to be contradictory.

1. Space (which is no object, no palpable thing, but merely the possibility of motion in every direction) is infinite. Yet the world, although immeasurable to us consists of a definite amount of matter and energy which can neither increase nor decrease.

2. The soul is a compound of highest complexity and is therefore destructible; but being a compound of a special form, it can be broken and built again. When built again, it can be improved. Souls of a special kind can be formed, and ever nobler ideas can be im-

planted into souls. Thus the soul—a special compound of living thoughts, living in the organized brain-substance of bodily beings as real nerve-structures—can continue to exist even beyond the death of the single individual; it can be propagated, transplanted, and evolved. And to accomplish this is the main object of human institutions. There is no immortality of the ego beyond the clouds, but there is a continuance of soul-life in this world. The continuance and higher development of soul-life is of vital importance, and the duties of our present lives must be performed, not to please or benefit ourselves but in a spirit such as to enhance the life of the race to come. We must live so that our soul shall continue to live and to evolve in future generations.

3. Freedom and necessity are not incompatible;* but freedom and compulsion are contradictions. If a man is compelled by the authorities of the law to observe the law he cannot be said to be free. But if the law— the good will to live according to the law and the honest intention to act with righteousness—is a part of the man and a feature of his character, he is free while observing the law. The actions of a moral man are necessarily moral; they are the necessary outcome of his free will.

4. The anthropomorphic idea of God as a transcendent personality is undoubtedly a paralogism of pure reason; but the conception of an immanent God as the cosmical law to which we have to conform in order to live and to continue to live in future generations is no paralogism, no logical fallacy. Such a conception of God is at variance neither with reason nor experience, and there is no atheist who could not be converted to

---

* See the writer's "Fundamental Problems," pp. 191–196.

it by rational argument and by a study of nature. This God is not the personified weakness of a benevolent father—the ideal of the deists who would fain make him as sentimental and feeble as they were themselves. This God is the stern severity of order and law—irrefragable and immutable as are all natural laws, and yet at the same time as reliable and as grand, as sure and eternal—visiting the iniquity of the fathers upon the children unto the third and fourth generation, and showing mercy unto the thousands of those that keep his commandments.

We thus have the three postulates of Kant again, although in another shape. We have no transcendental God, no illusory ghost-immortality, no freedom that stands in contradiction to the law of causation. But we have the immanent God of a moral law in nature; we have the immanent immortality of a continuance of our soul-life beyond death and the moral freedom of responsibility for our actions. The errors that were attached to these ideas are done away with, but their ethical value remains unimpaired. They have ceased to be postulates and have become truths—for now they are no longer paralogisms, they are free from contradictions ; they are real, because they represent certain facts of reality which can be verified by experience.

PROMETHEUS AND THE FATE OF ZEUS.

The Greeks possessed an old myth which in philosophical depth somewhat resembles the Teutonic Faust. The story of Prometheus is told in different versions by Hesiod in his "Theogony" (511 et seqq.) and in his "Works and Days" (48 et seqq.). Aeschylus, the first of the three great Athenian dramatists, gave in his great trilogy of the Fire-bringer Prometheus, the Bound Prometheus, and the Liberated Prometheus a third and undoubtedly the best, the most philosophical, and the profoundest version of the legend. And since these three great dramas exist only in fragments which bear witness to the grandeur of the Greek poet's thought, this greatest of all ideas, that of aspiring and conquering man—conquering through forethought—still awaits a great poet to give it a modern form. As Goethe created the final conception of the Faust-myth, so the poet of the future, perhaps still unborn, will let us have the final conception of the Prometheus legend.

Prometheus is the son of Themis, and Themis is the Goddess of law. Prometheus with the help of the eternal laws of existence has acquired the faculty of forethought. Prometheus means the man who thinks in advance.

Prometheus had a brother and his name was Epimetheus, that is the man who thinks afterwards, when it is too late. There is a story about an old Gotham magistrate who had very wise thoughts, but they did not come to him until the session was over and all the foolish motions of the fathers of the town had passed. His best thoughts came when he walked down stairs in the city hall. This same kind of wisdom, the wisdom of the staircase, was the wisdom of Epimetheus, and thus the two brothers were very unlike each other.

In those days Zeus kept the fire for himself; he allowed the sun to shine upon the earth and when he grew angry he threw down his thunderbolts upon oaks and mountain-tops. But he was envious and feared that man might become too powerful. Prometheus foresaw the great advantages which the usage of fire would have for mankind. So he stole the fire from the heavens and brought it to the people on earth, teaching them how to build a hearth and to use it wisely. But Zeus' punished Prometheus severely for his theft, he chained him to a rock and had an eagle swoop down upon him daily to devour his liver which always grew again during the night. Prometheus was afterwards liberated by the skill and courage of another daring man—by Hercules who shot the eagle and rescued the sufferer.

Why did Zeus not kill Prometheus? First we are told that Prometheus was immortal. But there is another reason still. Prometheus knew a secret which Zeus did not foresee, although it foreboded evil to the father of the gods. This secret, as we can surmise for several reasons, consisted according to the old mythological tradition in this: Zeus loved a goddess;

her name was Thetis, and it was written in the books of fate that the son of Thetis should be greater, infinitely greater, than his father. According to the version of Aeschylus, Zeus became reconciled with Prometheus on the condition that he should reveal the fatal secret to him so that he might protect himself against the imminent evil. And we are told that Zeus resigned his love and ordered Thetis to be married to a mortal man whose name was Peleus, and the son of Peleus was the greatest hero of Greek antiquity, the noble, the brave, the proud Achilles.

This is the version of Aeschylus, but there is another version still left. That is the version of the poet of the future. Aeschylus believes that Zeus was saved. Zeus being reconciled with Prometheus knew of the danger and evaded it. Yet we now know, that he could not evade it. Let a god have a son and the son will be greater than the god, even though the son of God may call himself the son of man. Says Goethe: "The son shall be greater than the father,"—that is the law of evolution, the law of life, the law of progress. We now know that Zeus was actually dethroned by a greater God than himself and this greater God was the son of man—the aspiring, the suffering, the conquering son of man.

Zeus is dead, but Prometheus is still living. Who is Zeus and where is Zeus? Zeus is the phantom-god of pagan antiquity. Zeus is a personification of the Divine in nature, he is a grand picture of God, but he is not God himself. If we expect that the picture we have made of God is God himself, if we imagine him to be a mind like ourselves, we shall fall into the same errors and pass through the same disappointments as did Prometheus. Says Goethe's Prometheus:

> "While yet a child
> And ignorant of life,
> I turned my wandering gaze
> Up toward the sun, as if above
> There were an ear to hear my wailings
> A heart like mine
> To feel compassion for distress."

It was most likely necessary that Prometheus should pass through his errors to arrive at truth, it was indispensable to brave the evils of life and to undergo severe sufferings in order to conquer. The errors as well as the sufferings, the very evils of life are good in so far as they help man to struggle and to progress. But in order to gain the victory, Prometheus ought to know that he must fight himself; he cannot rely upon the help of his phantom-god—of a Zeus above the clouds. The real God of nature is deaf to the prayers of those who pray in the hope that he will do the work for them.

There is more divinity in Prometheus than in Zeus. The God of the present time is the son of man and his symbol is the cross, which means that the way of suffering is the way of salvation, struggle is the condition of victory, the path of toil only is the road to a higher existence, the narrow gate leadeth unto life. The Zeus-idea of God is doomed and an infinitely greater, because truer, idea of God is dawning upon mankind. There is truth in mythology and there is a meaning in parables, yet the parable is told for the sake of its meaning and the truth is greater than mythology. Let us not be satisfied with mythology, but let us look out for the truth.

ENTER INTO NIRVANA.

THE RELIGION OF A FORERUNNER OF CHRIST.

THE religion of Buddha hinges upon the two ideas Sansara and Nirvana.

Sansara is the bustle of the world; it is full not only of vanity, but also of pain and misery; it consists of the many little trivialities that make up life. It is the pursuit of happiness; it is hunting for a shadow which the more eagerly it is pursued the quicker it flies.

The worldling lives in Sansara. He imagines he proceeds onward in a straight line, yet he moves in a narrow circle without being aware of it. He hastens from desire to pleasure, from pleasure to satiety and thence back to desire.

The worldling eagerly tastes the pleasure, and if he can he tastes it to the last, he intoxicates himself with it, only to find out that it was not what he had hoped for. Pleasure if tasted to the last becomes stale; it becomes staler than its symbol, the nectar of the grape that has been left in the glasses of topers after a night's carousal.

What is the result of a life in Sansara? Man's feet will become sore and his heart will be full of disappointment. The Buddhist says: The circular path of the Sansara is strewn all over with fiery coals.

Desire burns like a flame and satiety fills the soul with disgust. Enjoyment, however, is the oscillation between both. Desire is want; it is parching thirst and pinching hunger. It is destitution, poverty, dearth. Satiety, on the other hand, is not at all a preferable state. It is tedious and wearisome monotony; it is life without a purpose. The fulfilment of want means an emptiness of aspirations, it produces the nausea of maudlin misery, and the absence of desire is felt as an actual torture. A longing rises in the heart for the thirst of an unsatisfied desire and thus the pendulum swings back to the place from whence it came.

And happiness! What is the happiness of a worldling? It is merely an imaginary line between both extremes. The pendulum that swings to a certain height on the one side will necessarily reach exactly the same height on the other. It does not come to rest in the middle. There is no escape from this law, and if a man of the world be prudent he will moderate the oscillations so as to diminish the misery. Not going to the highest pitch of desire, he will not be obliged to drain the cup of myrrh to the lees.

Why does mankind continue to move in the circular course upon the fiery coals of Sansara? Because their eyes are covered with the veil of Maya. Individual existence, the Buddhists say, is a sham, an illusion, a dream woven of the subtle stuff of sensations. Man imagines that his sensory world is a reality. Buddhism teaches that the world of the senses is like a veil upon our eyes.

The veil of Maya does not exactly deceive man; on the contrary, the veil is the means by which man knows whatsoever he knows of truth. If the veil were not upon man's eyes, he would see nothing, he would be blinded, as was Moses in the presence of God. In itself the world of sensations is not a deceit, if it is not made so by being misunderstood.

The error, it is true, is natural. All errors originate according to natural laws; so did, for instance, the ideas of the flatness of the earth and of the apparent motions of the heavenly bodies. But if we err, the fault is not with the facts that lead astray, but with us. We deceive ourselves by our own error. The veil of Maya makes us feel our own being in contradistinction to that of all-existence; and this "we," the "I," the ego in its separateness is a self-deception. We live the dream of a pseudo-existence.

From the growth of the ego rise the self-seeking yearnings. Egoism begets egotism, and passions are the fruits of egotism. Passions produce pain and bring upon man the many evils of his earthly miseries.

Is there no escape from Sansara? Yes there is! The illusion that considers individual being as a reality can be destroyed. The veil of Maya can be lifted; which means, that its nature can be understood. In this way shall we recognize the error of egoism. There is no ego in the sense of a separate and individual existence, and with this truth it will dawn upon us that the regulation of action, as if there were an ego, is a fatal mistake. This mistake lies at the bottom of all the wretchedness of Sansara, and we can free ourselves only, so teaches Buddhism, by enlightenment, by understanding the truth, by abandoning the illusion. He who has attained enlightenment is a Buddha. Buddha means the enlightened one.

The highest stage of Buddhist perfection, the stage where a man becomes a Buddha is called Nirvana. Nirvana means extinction. As a flame is extinguished and ceases to be, so the *ignis fatuus* of the ego can also be extinguished. The egoistic error being extinguished, we enter Nirvana.

Nirvana means peace; it means liberation from illusion, and thus it brings a freedom of desire.

Nirvana is not annihilation. It is the annihilation of error only; and in this respect it reveals to him who lives in Nirvana, the higher life of true reality. In Buddhistic literature Nirvana is sometimes characterized in its negative aspect as an extinction of sham-existence, and sometimes again in its positive aspect as the life of truth and immortality. It is often described in most positive terms as true happiness, as a state of perfect bliss, as living in the realm of eternity, where there is no pain, no misery, no death. This appears to be contradictory to its literal meaning, but it seems to me that it is not.

As soon as we recognize the error of individual existence, we lift ourselves above the narrowness of egoism. We can in this state of mind contemplate our own fate from a higher standpoint; we can easily and we do willingly give up our pursuit of happiness; we can live in this world as though we were not living. Our "we," our "I," our "ego," the separateness of our individuality has ceased to be, and the life of the universe lives in us. We have become stewards of cosmic existence. In this way our joys as well as our pains are transfigured and a divine peace will inherit our souls that are now free from desire.

Pain, together with the vanity of pleasure, will diminish in the degree of the enlightenment attained. This is a law that is demonstratable in such exact sciences as physiology and biology. Our scientists inform us that the use of the sensory nerves blunts feeling and favors intellection. The highest sensory nerve, in which the intellectual element is comparatively most perfect, is the optic nerve. The retina of the optic nerve, while perceiving the differences of infinitesimally small fractions in ether-waves, has become insensible to pleasurable as well as to painful feelings.

The idea of Nirvana, it must be said, is of a most dangerous character, if it is conceived as mere pessimism in its negative features alone. It will in that case lead to apathy, to destruction and death. Did perhaps Gautama Buddha himself conceive Nirvana in a spirit of negativism? Perhaps he did. At least it is certain that many of his disciples did; for the Buddhism of the East has produced most fatal effects of indifference and retrogression upon those races that embraced its faith.

If Nirvana is conceived in its negativeness, Buddhism will be a dualistic religion. In that case we have existence and non-existence, Sansara and Nirvana, sham-reality and nothingness. If, however, Sansara is conceived as an illusion and Nirvana as the destruction of the illusion, we need not resort to the nihilistic world conception of a dual nothingness; we need not derive from the Buddhistic premises the negative ethics of destroying life together with the illusion of egoism.

One of the most important truths proclaimed by Buddha, was the doctrine that man can enter into Nirvana while he lives. When Gautama had found redemption from the evils of existence, he resolved to announce his gospel to the world. He went to Benares and on the way he met one of his old acquaintances who asked him:

"What is it that makes you so glad and yet so calm?"

Buddha answered:

"I have found the path of peace, and am now free from all desires."

Little interested in Gautama's bliss, his acquaintance further enquired where he was going; and we are told in the Buddhist legend:

The Enlightened one answered:

"I am now going to the city of Benares to establish the kingdom of righteousness, to give light to those enshrouded in darkness, and open the gate of immortality to men."

He gave up fasting, for he looked upon the oppression of the body as a vain effort of conquering the evils of existence. He abandoned asceticism as a means of salvation.

It seems strange that life can be gained only through annihilation of self; immortality is possible only through the death of the transient and the happiness of eternal peace will come with the crucifixion of the desire for happiness. It seems strange, but it is not. However, it is natural that the deeper a truth is, the more contradictory it will appear to those who are prisoners still in the bondage of error.

Buddha's doctrines were misunderstood, misinterpreted, and misused. Yet they have given strength in temptation, comfort in misery, peace in tribulation, solace in death to many millions of toiling, aspiring and suffering human hearts.

THE HUMAN SOUL.

The practical purpose of Religion is the salvation of human souls. When Jesus was walking by the Sea of Galilee, he saw two brethren, Simon, called Peter, and Andrew, his brother, casting a net into the sea: for they were fishers. And he sayeth unto them: "Follow me and I will make you fishers of men."

And so shall ministers be fishers of men to save human souls. But how can they save human souls when we are told that modern psychology is a psychology without a soul? The immortal soul, consisting of a trancendent substance, as it was supposed to be by the old schools of orthodox theology, does not exist. There is no such a thing as an eternal and mystical ego which continues to live even if the body dies.

The great Scotch philosopher, Hume, said:

"As for me, whenever I contemplate what is inmost in what I call my own self, I always come in contact with such or such special perception as of cold, heat, light or shadow, love or hate, pleasure or pain. I never come unawares upon my mind existing in a state void of perceptions: I never observe aught save perception..... If any one, after serious reflection and without prejudices, thinks he has any other idea of himself, I confess that I can reason no longer with him. The best I can say for him is that perhaps he is right no less than I, and that on this point our natures are essentially different. It is possible that he may perceive something simple and permanent which he calls himself, but as for me I am quite sure I possess no such principle."*

* Hume, Works, Vol. I, p. 321.

Modern psychology has fully adopted Hume's position. If a man speaks about himself, he means perhaps his body, or a certain part of his body. In another case he may mean a special idea of his mind. It is that idea which at the time is prominent in his soul and which he pronounces as his opinion. Formerly it was supposed that the ego who pronounces the opinion, "I say this and I say that," was one thing, and the opinion adopted by that ego another thing. And this ego was supposed to form the basis of man's personality, its supernatural unity. Modern psychology now contends that this ego is identical with its opinion; the "I say" is identical with the idea pronounced; or, in other words, the ego and its contents are one. Accordingly, our ego is constantly changing, for it is now this, now that idea, which is prominent in our mind. An ego by itself, a thinking subject without an idea, a perception, or sensation to be thought or felt, does not exist; and our soul is nothing but the sum total of all the ideas that live in our brain.

Very well then! The ego, as a thing independent of its contents, is a sham and always was a sham. Can a man be afraid of losing that which he never possessed? Certainly not! Renounce that ego, and abandon your anxiety about its preservation.

The matter, however, is different concerning the preservation of your soul. Is the soul of man less valuable since it has been proved that it lacks the unity which the ego was supposed to afford to it. Not in the least! Our soul, whatever it be, remains as valuable and precious as ever; if it is the sum total of our thoughts merely, yet that is the sum total of our intellectual and moral existence. The ideas which in their totality constitute ourselves, are the elements that condition our actions, and our actions shape our future ; they will lead us to higher planes, or they will undo us and wreck our lives. Thus the purpose of religion, to save the souls of man, it seems, becomes rather more urgent than before.

Science seems to destroy Religion. But it does not ; it destroys its errors only. Indeed, it becomes the basis of Religion, and Religion based on Science will be truer, purer, and grander. When our long cherished errors fade away before the light of science, life appears so empty and truth seems void of comfort. Let us, however, not be dismayed! After all, truth is better than error and a deeper insight always proves in the end that the truths taught by science are by far nobler and greater than the loftiest fiction and fairy-tales of our imagination can be.

Let us but consider how easily souls are lost! The purpose of religion becomes more imperative when we bear in mind the fact that souls can grow and expand. We can implant in the souls of men new thoughts and purer ideas, which will preserve them in temptations and guide them through the many allurements and dangers of the world. We can by instruction and example transmit to the minds of children our own souls, and thus build again our characters in the growing generation.

Gœthe said : 'The son should be better than the father'; and yet the son can be better only if the father rears the better part of himself in the soul of his son. Thus humanity will progress, it will advance more and more in the triumphant march of evolution.

The child that lies in the cradle possesses a most precious soul. But the infant's mind is a promise rather than a real and full grown soul; it is the bud not the fruit; it is a dear hope, a potentiality, but not the harvest of maturity. The most valuable parts of his soul, the elements of manly strength and of moral character that will give stability to his will and direction to his purpose, must be implanted into the tender mind of the child. All that which makes of man a human being, must be grafted upon the inherited predisposition of his mind. And how easily is that purity lost, how quickly is that innocence gone which appears as the sweetest charm in the beaming eyes of children. When they come in contact with the lower tendencies of life, how readily evil thoughts enter their minds and impure ideas poison their souls and the habits of their lives. Therefore David prayed, "Create in me a clean heart, O God, and renew a right spirit within me."

The soul of man is not immortal in the sense that matter and energy are now known to be indestructible. On the contrary, the soul of man *is* mortal. But seeing that we can make it immortal, that we can preserve our souls even after death in the coming generations, that we can implant our spirit in our children, the purpose of religion grows in its scope and importance.

The pure, the noble, the great, the moral thought will live and will exercise upon everyone a wholesome influence. If our soul is the sum total of our hopes and wishes, of our aspirations and longings, of our concepts and our ideas, let us take heed and beware not to introduce evil thoughts, but let us receive into our soul the love of truth and the eagerness of performing that which is right and just. This is the object of religion; and this is the sum total of the religion of mankind.

Bad thoughts as well as good thoughts are like leaven; and a little leaven leaveneth the whole lump. Let the religion of humanity thus enter your souls not as words without a meaning that slumber like a dead letter in your mind, but as a power of enthusiasm which, like the leaven in meal, pervades all your thoughts and sentiments until the whole be leavened and changed into another and better substance.

Morality is that which preserves our soul, and it is the moral part of our soul only which we wish to preserve in our children. Immorality is that which leads to wreck and ruin, but morality makes life everlasting.

The immortality thus acquired is greatly different from the old dogmatic view of immortality. The old view of immortality is a chimera of Utopian character, the new view is a truth established by science, a truth that can be verified. The old view of immortality is a holy legend, and the best that can be said of it is this, that it foreboded the true view of immortality which teaches that there is a continuation of our soul-life after death. This continuation, however, is not an inherent quality of the soul, nor is it given to us as an act of mercy. The continuation of our soul-life must be acquired by our own efforts, it must be worked for, it must be earned by hard struggles, and it must be deserved.

I see all the world gathering earthly treasures to leave an inheritance to their children. But I see few who care for their souls. All interests are taken up with the desire for riches, but the most valuable riches which you might possess, remain neglected. You provide for meat and raiment and other necessities of life, but you disregard to provide for the immortality of your soul.

What are all the possessions of man if he is not wise enough to use them well, and what is power and earthly blessings if the men to whose lot they have fallen, cease to progress or even commence to degenerate? What is an inheritance left to your children, be it ever so great, if you disregard the education of their souls? The word of Christ will forever remain a great truth: "What shall it profit a man, if he shall gain the whole world and lose his own soul?" (MARK VIII. 36.)

THE UNITY OF THE SOUL.

The main difference that obtains between the old and the new psychology concerns the unity of the soul. The old psychology considers the soul as an indivisible being whose centre is found in the ego. This ego-entity is said to be the subject of the psychical states; it is the subject in the original sense of the word; i. e. that which underlies. The soul, according to this view, is not the feelings and the thoughts which ensoul a human being, but it is a mysterious something which is in possession of feelings and thoughts, and the nature of the mysterious something, of the underlying subject, is unknown to us.

Modern psychology does not consider the soul as an indivisible being. The soul is not an ego-entity, a subject, that has feelings and ideas, but these feelings and ideas are actual parts of the soul. A man's soul is the totality of his feelings, of his thoughts, of his ideals.

This view may easily and wrongly be interpreted as if the soul were simply a heap of feelings, as if no unity existed and as if the ideas dwelling together in one and the same brain were like a bag of peas, which have no connection, no bond of union, among themselves. This is not so. The feelings, ideas, and ideals in a man form indeed a unity—only this unity is a hierarchical system, it is a unity of arrangement and does not mean that the soul is an indivisible unit or a kind of psychic atom. This truth can most clearly be expressed by contrasting the two views in two German words: The soul is not an *Einheit*, but an *Einheitlichkeit;* not a *unit*, but a *unification*.

And the unity of the soul produced by unification is by no means an indifferent quality. The unity of the soul, I feel almost constrained to say, is the soul of the soul. The way in which certain ideas are combined in a unity constitutes the most individual and most remarkable and also the most characteristic feature of a personality. Also the energy of nerve-action, the vigor with which the different ideas respond to their stimuli is of incalculable importance.

Suppose we could put together the soul of a man from a given number of ideas as we put together a mosaic from a given number of colored stones. The stones and their colors, their brightness, their shape and the variety of their colors are of importance, but the pattern will after all make the picture of the mosaic. The same ideas are put into the minds of thirty or forty children in one and the same class-room, but how differently do their minds develop! Even children of the same parents who live in the same surroundings and under the same conditions, receiving the same instruction and having before their eyes the same examples, will develop quite distinct and divergent individualities. The very same thoughts in two different minds do not necessarily produce a sameness of soul. In one mind everything may be methodically arranged, so that on the proper occasion the proper thoughts turn up at once and all the ideas form a system, so that order reigns everywhere. Again in another mind there may be the very same thought-material, yet order is lacking, confusion prevails, everything stands topsy-turvy as if the brain were an old lumber-room in which things have been set aside without any plan of consideration.

It is wonderful how rich the possibilities of soul-patterns, so to speak, are! We cannot say that this one and this one only is the true ideal soul, for, provided that those indispensable soul-structures which constitute the humanity of a man are not lacking, we may have and indeed we do have, an unlimited variety of personalities, the beauties of each being peculiar to themselves.

People often show a tendency to classify the personalities of great men in higher and lower classes asking such questions as these: Who was greater Shakespeare or Goethe? Plato or Aristotle? Bismark or Moltke? The answer is, we cannot measure the greatness of mind by a scale so as to have the great men of thought and action classified by degrees as number one, two, three, etc.

The soul of man, being the organisation of his ideas, is too subtle a substance,—indeed we should not even call it so for it is form and not substance—the soul of man is too subtle to be weighed or measured, and the worth of a noble soul is so peculiar, so unique that, irrespective of its shortcomings which we must expect even great men to have, we can compare one soul with other souls only in order to set them off by contrast and to appreciate their qualities by contrast, but we must recognise that each soul possesses a special charm of its own, each soul is an individuality which as such is not classifiable as higher or lower, better or worse than other individualities.

Individuality being a natural and also a most valuable feature of a man's soul, it is our duty to respect individuality. Every man has a right to be individual provided the traits of his individuality do not come in conflict with the rights of his fellow-beings. And the application of this right in educational affairs is greater still. We are bound to respect the individualities of children also. Parents, educators, and teachers have to observe and study the charâcters of the souls entrusted to their care. They have to prune and guide the growth of individualities wherever whims and vagaries arise, yet they should do so with due discrimination and with a becoming respect for the individuality of the growing minds.

GHOSTS.

The Norwegian poet Henrik Ibsen has written a most awe-inspiring drama under the mysterious title "Ghosts." Does this most modern author believe in spirits? Does he take us into a haunted house? Are not ghosts and haunted houses left as a survival only? O no! The ghosts of which Henrik Ibsen speaks are everywhere; they are not exceptional cases; for we ourselves are visited by the spirits of former ages; our brain is haunted by ghosts. It is full of the proclivities, the dispositions, the ideas, and the sins of our ancestors.

Mrs. Alving, the widow of a dissolute husband, and mother of a son whose life has been poisoned by his father's sin, witnesses her son's behavior in the adjoining room. It is the exact repetition of a scene in which her husband had played her son's rôle some twenty years ago. There is his ghost reappearing. In considering the weighty seriousness of the truth, that we inherit, not only the character of our ancestors, but also the curses of their sins; that all our institutions and habits are full of ideas inherited from a dead past, she says: "I am afraid of myself, because there is in me something of a ghost-like inherited tendency of which I can never free myself. . . . . I almost think we are all of us ghosts. It is not only what we have inherited from father and mother that reappears in us, it is all kinds of dead old beliefs and things of that sort. These ghosts are not the living substance of our brain, but they are there nevertheless and we cannot get rid of them. When I take up a newspaper to read, it is as though I saw ghosts speaking in between the lines. There must be ghosts all over the country. They must be as thick as the sands of the sea."

It is perfectly and literally true that our soul is haunted by ghosts; nay, our entire soul consists of ghosts. Our brain is the trysting place where they meet and live; where they grow and combine, and in their combinations they propagate, they create new thoughts which according to their nature will be beneficent or baneful.

What are these ghosts? They are our experiences, the impressions of our surroundings upon the sentient living substance of our existence. They are the reactions that take place upon the impressions of our surroundings; they are our yearnings and cravings; they are our thoughts and imaginations. They are our errors and vices, our hopes and our ideals.

Henrik Ibsen shows that the ghosts which are the inherited sins of our fathers lead unto death. What an overwhelming and horrific scene is the end of the drama, where the son asks his mother to hand him the poison in case the awful disease will pass upon him which will soften his brain and spread the eternal night of imbecility over his soul. The mother in her anxiety to calm her son's wild fancies, promises to do so: "Here is my hand upon it," she says, with a trembling voice: "I will—if it becomes necessary.

But it will not become necessary. No, no! It will never become a possibility."

There is a law of the conservation of matter and energy; but there is also a law of the conservation of the stuff that ghosts are made of. The law holds good not only in the material world, but in the spiritual world also. Every vice transmits its curse; and the moment comes when the unfortunate mother has to face the fatal attack of the terrible disease.

The heroine of the drama, the innocent and wretched mother had sought help of the clergyman—the man whom she had loved. When her husband had betrayed her, had poisoned her in her youth, she fled to him in wild excitement and cried: "Here I am, take me!" But the clergyman's stern virtue had turned her away from his door, and he prevailed upon her to remain a dutiful wife to her vicious husband. She had tried to find comfort in the religious injunctions which he preached to her. She lived a life in obedience to what he represented as her duty. But now she says to him: "I began to examine your teaching in the seams. I only wished to undo a single stitch, but when I had got that undone, the whole thing came to pieces, and then I found that it was all chain-stitch sewing-machine work."

The distressed woman feels only the curse of law and order which have been invented for the salvation of mankind. Her experience leads her to trust rather in anarchy than in the threadbare superstition which our generation has in favor of the letter of the law. The sternness of virtue cannot save us, nor our blind obedience to sanctified traditions. She exclaims: "What nonsense all that is about law and order. I often think it is that which exactly causes all the miseries there are in the world. I can no longer endure these bonds; I cannot! I must work my way out to freedom!"

Here lies the cure of the disease. We must work our way out to freedom. The simple method of shaking off law and order will only increase our troubles. We must learn to understand the nature of ourselves. By patient work alone can we exorcise the evil spirits that haunt our souls; and we can nourish and foster those other spirits which shower blessings upon our lives and the lives of our children. We cannot escape the natural law which, inviolate, regulates the growth of our souls; but we can accommodate ourselves to the law and the same law, that works disaster and death, will produce happiness and life.

Superabundance of life gives a power that might produce great and noble results. But when the life is stagnant as was that of Mrs. Alving's husband, a vigorous youth exuberant in strength and health, an unsatisfiable craving for pleasure takes the place of a want of activity; and instead of useful work, vicious habits are produced. The germ of many diseases is a morbid pursuit of enjoyment. Pleasure is made the aim of life, leading astray step by step into the abyss of misery and death. Not that happiness and pleasures were wrong! But it is wrong to make of them the purpose of life. Let happiness be the accompaniment of the performance of duty and happiness will follow as the shadow follows the body. If we pursue happiness, we turn our back upon the sun of life and we shall never find either satisfaction or happiness.

* * *

The law of the conservation of soul-life with its blessings and its curses has not only a gloomy side, it has also a bright side, and it behooves us when considering our heir-loom of curses, to remember that they are small in comparison to the grand inheritance of blessings which have come to us from thousands of generations. What is all our activity, our doing, and achieving, our dearest ideals—what are they but the torch of life handed down from our ancestors? Gustav Freytag, the German novelist, might also have called almost all his novels "Ghosts." Especially the "Lost Manuscript" and the series of novels called "The Ancestors" are studies illustrative of the same truth. Yet while Ibsen paints the dark side only of the law of the conservation of ideas, Gustav Freytag paints the dark *and the bright* sides. Gustav Freytag says:

> "It is well that from us men usually remains concealed, what is inheritance from the remote past, and what the independent acquisition of our own existence; since our life would become full of anxiety and misery, if we, as continuations of the people of the past, had perpetually to reckon with the blessings and curses which former times leave hanging over the problems of our own existence. But it is indeed a joyous labor, at times, by a retrospective glance into the past, to bring into fullest consciousness the fact that many of our successes and achievements have only been made possible through the possessions that have come to us from the lives of our parents, and through that also which the previous ancestral life of our family has accomplished and produced for us."

We have to bear the evil consequences of the vices of our ancestors, but these evils can be overcome; and when they cannot be overcome, they will after all find a termination, for death is the wages of sin.

The nature of sin is its contrariness to life; its main feature is the impossibility of a continued existence. Extinction being the natural consequence of viciousness, the wages of sin are at the same time the saviour, the redeemer from the evils of sin.

If all the parents in the whole world were like Chamberlain Alving, the ruthless father of Oswald Alving, and like Mrs. Engstrand, the frivolous mother of the coquettish girl Regina, humanity would soon come to an end. It may be that none of us is entirely free from these traits; but some of us are so more or less. In some of us these traits are mixed with ennobling features, and we are striving to overcome that which we have recognised as bad. However, nature is constantly at work to prune the growing generations. Death is the wages of sin, and the bright side of this awful truth is the constant amelioration of the race.

THE RELIGION OF RESIGNATION.

Among the many religions upon earth there are two that exceed all the others in the number of their devotees. They are Buddhism and Christianity. Neither Judaism nor Mohammedanism, nor even Paganism, can approach them. The latter taken together do not as yet equal one of the two former; it is as if the world were divided between them.

Buddhism and Christianity have one common feature. Both proclaim the gospel of salvation from the evils of this world by resignation. Both point to a higher life which can be gained through the sacrifice of our individual selves. Other religions require sacrifices of lambs and goats: Buddhism and Christianity demand the surrender of self. Mohammedanism promises enjoyment and happiness upon earth and in heaven. Both Buddhism and Christianity preach endurance in affliction and submission to tribulation.

It seems natural to seek pleasure and to shun pain. The religious injunctions of Buddha and Christ are a reversion of this instinctive desire. They preach this: Do not shun pain, and do not seek pleasure. Says Christ: "If any man will come after me, let him deny himself, and take up his cross and follow me."

Buddhism, it is well known, did not succeed in overcoming the many superstitions of its converts. Christianity became ossified as soon as its mythology was systematized into theological dogmas. The Christians, while clinging to the letter of their creed which killeth, lost the spirit of their teacher's doctrine.

We shall not here point out the errors of these religions, but try to find the key to their wonderful success. And can there be any doubt about it?

Their success can be due only to their thorough conquest of death. The Buddhist who has taken his refuge in Buddha, and the Christian who is earnest in his following of Christ, will not tremble in the face of death, for to them death has lost its sting. The Christian lives in God, and the Buddhist has even upon earth spiritually entered into Nirvana. They have placed all their hopes in a higher life, "where there shall be no more death, neither sorrow nor crying, neither shall there be any more pain, for the former things are passed away."

This victory over death is not accomplished by avoiding death and by shunning the anguish of life, but by a surrender to death of that which cannot escape death and by finding rest in the ideal world of immortal life. Whatever be our fate,—they say unto themselves,—the kingdom of God will be victorious; all other things are mere trifles; therefore let us remain children of God and we shall inherit his kingdom. Luther sings:

> Strong tower and refuge is our God,
> Right goodly shield and weapon;
> He helps us free in every need,
> That hath us now o'ertaken.
> Take they then our life,
> Wealth, honor, child and wife,
> Let these all be gone,
> No triumph have they won.
> The kingdom ours remaineth.

This song with its powerful melody was the slogan of the new faith that regenerated Christianity and conciliated religion with the progress that science had made before the Reformation. Yet Luther and other Christians believed in the immortality of their ego,' and it seems as if their religious confidence were based upon this error. We have ceased to believe in a mystical soul-substance which was formerly supposed to inhabit the body as a stranger, and which after death will hover about somewhere as a spectre. We have ceased to believe in ghosts; science has banished the phantoms of disembodied spirits out of the provinces of psychology and philosophy. But must we for that reason cease to believe in life and in spiritual life? Must we therefore consider death as a finality? Does not science teach the persistence of life and of spiritual life; and is there the slightest reason that we should cease to believe in the immortality of our ideals? Is it not a fact, scientifically indubitable, that every work done, be it good or evil, continues in its effects upon future events? Is it not a fact established upon reliable observations that the evolution of mankind, and of all life generally upon earth, is one great and continous whole; that even to-day the efforts of our ancestors are preserved in the present generation; their features, their characters, their souls now live in us. Certainly not all features are preserved, but those only which nature considered worth preserving. So our characters, our thoughts, our aspirations, our souls will live in future generations, if they are strong enough, if they are noble and elevating. In order to be strong, they must be in accord with nature, they must be true. In order to live, they must be engendered by the evolutionary tendency in nature, which constantly endeavors to lift life to higher planes. It must be, as the Christian expresses it, in harmony with

God, if God is meant to be that power in nature and in our hearts that ever again and again prompts us to struggle and to strive for something higher.

Our soul can no longer be considered as that unity which it used to be to our forefathers. It is a part of the soul of humanity in a certain phase of its development. As such it is a rich combination of certain ideals, thoughts, and aspirations of hopes and fears, of wishes and of ideals. Our ego is nothing but an ideal thread on which are strung the invaluable pearls of our spiritual existence. The ego is nothing but the temporal succession in which these ideas are thought.

It is not the belief in an immortalized ego that can conquer death, but it is the surrender of this ego and of all its egotistic desires. This ego we now know is no real thing ; it is an illusion and possesses a fleeting, momentary, sham existence only. Reality of life is not to be found there, and if its continuity is broken in death, our individual existence ceases, but not necessarily the life of our soul. The ideal world of our mind can outlive our body, and we can gain an immortality of that part of ourselves which is most worthy of being preserved.

This it appears, is the truth in Buddhism and Christianity, this is the secret that explains why they conquered the world. Resign all egotism, do not place your hope upon this fleeting existence, and devote your efforts to the creation of that higher life, of that ideal world, where death is unknown and the petty tribulations of life disappear !

This life cannot be realized by the poet and philosopher only, not by the great only, the heroes of mankind : it can be realized by every one of us. It is this that Christ preached, and it is this that Buddha proclaimed. Every one of us is called to participate in the higher life, for the intellectuality of a higher life is one phase of it only, and it is not its grandest part. Its sum-total is comprised in all those many ideals and aspirations that, in one word, we call morality. It is, as Paul says, Faith, Hope, and Charity; but Charity is the greatest among them.

Men who have given up their individual ego, who have risen to the height of that spiritual life which knoweth not death, will live in this world as though they lived not; they that weep, as though they wept not, and they that rejoice as though they rejoiced not, and they that buy as though they possessed not, and they that use this world as not abusing it: for the fashion of this world passeth away.

They will live, as though they lived not, because their life is no longer the fleeting sham-existence of their egotistic desires. Their life has become an expression of that higher life which is immortal. They buy as though they possessed not, because they know that they shall have to leave their possessions.

They consider themselves as stewards to whom property is entrusted for a wise use. Even their joys and pains, their recreations and troubles become transfigured by the universality of the spirit that animates their whole being.

The religion of the future will not be Christian dogmatism, it will be no creed, no belief in any of the tenets of the church. Yet it must preserve the spirit of Christianity which has enabled it to conquer death. It must be a religion of resignation. If thou wilt enter into life eternal, cease to cling to that which perishes, and become one with the Life Immortal!

THE RELIGION OF JOY.

The Christian gospel is a tiding of joy; but its joy is very different from the happiness that is so eagerly sought for by thousands and millions of wretched beings who tire themselves out by hunting shadows.

It is natural that only two religions have a festival of rejoicing in the birth of a child destined to be the saviour of the world; Buddhism, namely, and Christianity.

Buddhism and Christianity are the religions of resignation. They demand that we shall willingly and unhesitatingly take up our cross; that we shall not shirk tribulations, suffering, and least of all death; that we shall renounce all cravings for pleasure, sacrifice all desires of egotism, and in fact give up our very self, which is the source of all our unsatisfied yearnings.

Buddhism and Christianity, being religions of self-denial, have been called pessimistic world-conceptions. In a certain sense they are pessimistic, in another sense they are not. They ought to be called melioristic. Recognizing to its full extent the truth of pessimism, recognizing all the misery that exists in the world and the wretchedness of living creatures, the religions of self-denial are preached to show the path of salvation. In this sense Buddhism and Christianity are the religions of joy.

Says the Apostle: "Rejoice always!" and again he describes himself and his co-workers as the ambassadors of Christ: "As unknown, and yet well known; as dying, and behold we live; as chastened, and not killed; as sorrowful, yet always rejoicing; as poor, yet making many rich; as having nothing, and yet possessing all things." Says Christ: "Rejoice and be exceeding glad!" and the angel said to the shepherds: "Fear not, for behold I bring you good tidings of great joy, which shall be to all people."

Wherever a religion of self-denial has been preached, it has always been a gospel of cheer, of gladness, of salvation. This seems to be contradictory, and yet it is natural.

The main idea of the religions of self-denial is a truth which, if lost, we should have to discover again. Similarly, if our knowledge of the law of gravitation were lost, we should have to discover it again. And if another than Newton had calculated its formula, the formula would be exactly the same as it is now, whether it were expressed in Greek, or in English, or in Chinese.

Spiritual truth is no less rigorous than mathematical truth. Spiritual truth has to develop according to law no less than the flowers in the fields, no less than human civilization, the arts and the sciences. When the blossoms blow in springtide, it appears as if the earth had long been preparing and expecting this moment. Thus when Buddha went under the fig tree, where the idea of salvation enlightened his mind, the Buddhistic gospels relate that the angels sung, "This is the night the ages waited for."

Is it surprising that so wonderful a truth as that life is love, salvation is self-surrender, and joy is the sacrifice of all desire, has been clothed in myths and decked with miraculous legends? Is it surprising that great institutions were founded with ceremonies and rites in order to make comprehensible this spiritual truth to those who could not grasp it? And again, is it surprising that in all these institutions the truth is overgrown and hidden by the myth? The letter that killeth has prevailed over the spirit!

Science with ruthless criticism destroys the mythology which has so long prided itself as the truth. Yet science will never destroy the truth which has been the vitality in the germs from which sprang Buddhism as well as Christianity. And the religion of science, if it is to be a live power, must preach the same truth.

Science recognizes the struggle for life, but the religion of science brings peace. It brings the peace of soul that makes man one with that power which is the source of all life, one with that actuality which is the way, the truth, and the life; so that what appeared as a struggle for selfish ends, now becomes work, and work, whatever it be, pleasant or disagreeable, sowing or reaping, ruling or obeying, drudgery or the work of enthusiasm and love, is all transfigured by being conceived as the performance of duty.

The religion of science does not preach asceticism, when it demands self-denial and a radical surrender of egotism. On the contrary, like the good tidings of Bethlehem, it proclaims a religion of joy—not for those who are rich, but for all the world; first for the poor, yet also for the rich, if their hearts are fit to receive the gospel.

THE FESTIVAL OF RESURRECTION.

Spring comes again; and Eastertide reminds us of nature's immortality. There is no death! What seems so is transition. When in wintry weather the sun hides his face, northern blasts tear the leaves from our trees; but now the sun is returned and new life grows on every branch, the verdure reappears in the fields and man's heart believes with strengthened confidence in the realization of human ideals.

Easter day is the festival of Christ's resurrection, and the question has often been raised whether Easter day can with any consistency be celebrated by those who have ceased to believe in the sacred legend that Jesus Christ who died on the cross rose on the third day from the dead. We firmly maintain that it can and that it ought to be celebrated by all those who believe in the revival of spring, in the constant resurrection of human life, and in the immortality of our ideals.

Eastertime is not at all exclusively a Christian festival; Eastertime is a festival of natural religion. Its very name is pagan, for *Ostara* was the goddess of the returning light; and light brings life. She was the *Aurora*, the *Eos*, of the Germans, the deity of the morning dawn in the East; and the egg was the holy symbol that represented her mysterious powers.

An egg is a wonderful thing; it has been the object

of repeated investigations by our greatest naturalists; and our profoundest philosophers have pondered over the revelations of its marvelous secrets. The egg represents the potentialities of life. Mere warmth is needed to change the apparently homogeneous and insensible yolk into a most complicated animal endowed with a certain degree of intelligence. The egg represents, as we now know, the actual memories of chicken-life up to date. Its memories are not conscious memories, but the preservations of certain structures in living matter. They are motions of a certain form, which under favorable conditions and proper temperature will repeat all those motions, those vital activities, which its innumerable ancestors went through in uncounted ages past.

How wonderful are the secrets of form, and, in spite of the complex applications of which the laws of form admit, how simple is the basic idea that explains their mysteries! The artillerist, who aims his cannon, knows that a hair-breadth's difference in the angle of elevation will give another course to the missile; the curve of its motion will be changed with the variation of its determining factors.

The egg contains the determining and formative factors of certain motions of living substance, not otherwise than three points represent the potentiality of a special kind of curve. The determining factors of the egg have, in their turn, been determined by the parental activities of its predecessors; and thus the egg becomes a symbol of resurrection.

Life is not extinct with the dissolution of individual existence, for even the individual features are preserved in coming generations. And, if this be true in the chicken, how much more is it true in man.

Man's intellectual life has still other channels to be preserved in and transmitted to the souls of other men. These channels are human speech. The spoken word, and perhaps more so, the written or printed word, make it possible for the valuable thoughts of great thinkers and the enthusiastic aspirations of poets to live among us as if their authors had never died. Indeed, they have not died, they live still. Their souls are, and will remain, active presences in mankind to shape the destinies, and to guide the future development of our race.

Whether any given one of the heroes of mankind rose bodily from the dead or not, especially whether Christ rose bodily from the dead or not, is quite indifferent for the truth of the constant resurrection which, as science teaches, continuously takes place in nature and in the evolution of humanity. Let us not say, because there is no truth in the fables of religious mythology, that there is no resurrection whatever. Let us not say that we do not care for such a resurrection as can be observed around us in nature, and as can be experienced in human soul life; that unless we rise as bodiless spirits, as taught by supernaturalistic religions, we do not care for any resurrection in which the continuity of our individual consciousness is interrupted. Let us not speak like spoiled children, who want their caprices fulfilled, and if they cannot have their whims satisfied, want nothing at all. Let us rather become familiar with the real facts of life, and we shall learn that truth is grander than fiction, and real nature is better than an imaginary supernature.

We are told by men that aspire to be radical freethinkers, that this conception of immortality is a revival of old superstitions. What a strange misconception! Man will die, they say, and if man is dead, all is over with him; death is an absolute finality; and no one, so they maintain, will care for any other than a personal immortality, in which the continuity of consciousness is preserved.

Men of this class are not familiar with the facts of life. Not only is it true that life continues after the death of the individual, and that the work of every individual continues as one of the factors in the formation of the destinies of future generations, but also the care for what will be the state of things after our death is a most important motive in all our actions. We do care for what will take place after our death. We do care for the fates of our children, of our nation, of our country, of our ideals and hopes, and how our soul-life will affect the future development of mankind. We do care for such a continuance after death, we do care for an immortality of ourselves, even if the continuity of our consciousness be broken. The fact that we care for such things is the basis of ethics; it makes of man a moral being. This is the motive that compels even those who do not believe in personal immortality, to sacrifice their lives for their beloved ones, for their convictions, and for their ideals.

Let us celebrate Eastertime as one of the most prominent festivals of natural religion. It is the feast of resurrection, it proclaims the immortality of life, and preaches the moral command, not to live for this limited life of our individual existence only, but to aspire to the beyond. Beyond the grave there is more life, and it is in our power to form and to shape that life for good or for evil.

THE CONQUEST OF DEATH.

JESUS CHRIST said to his disciples: "In the world ye shall have tribulation, but be of good cheer, I have overcome the world!"

This is the grandest advantage of religion that it comforts him who has religious faith, while he who has it not, must tremble in this world of worry, of turmoil, of struggle, and of death.

A scientist who had pondered over many deep problems and had been successful in the solution of several mysteries of nature, said with suppressed emotion: "Religion is a sweet self-delusion that helps us to overcome the desolateness of life."

Why is it a self-delusion? Because, he might have answered, the ground upon which religious comfort is based, is scientifically untenable; yet is it sweet, because religion alone can overcome the vanity of the world; religion alone can fill the emptiness of a perishable fleeting life that seems to consist only of troubles and cares, the joys of which, if closely examined, are found to be stale and unprofitable.

The Christians, it may be conceded, delude themselves when believing all the many dogmas of their church. But is it a self-delusion, if they have really conquered the world, and if they face all the agonies of death with equanimity? Granted that their belief is wrong, we often observe their moral courage to be of the right kind. They prove by their example that death can be conquered, that we can raise ourselves above the narrow sphere of selfishness and lead a life that is inspired by the religious ideal of a victory over death.

I confess that I am not a believer in the current doctrines of the Christian churches, but at the same time I openly declare, that I am a believer in Religion. I have no theological creed to which I adhere, I know of no confession of faith which I would adopt, but I have a faith, that man, without any act of self delusion, can overcome the desolateness of life; he can fill the emptiness of existence with imperishable treasures —with those treasures that are laid up in the spiritual empire of human aspirations, where neither moth nor rust doth corrupt, and where thieves do not break through nor steal. I have a faith, that man can conquer death and can build an ideal life of spiritual loftiness upon the material existence of his being.

This faith is that of the mustard-seed. This faith does not look behind as do all the creeds; this faith looks forward. This faith does not anxiously cleave to the past; as do all the dogmatic confessions of faith. The right kind of faith, the only faith which deserves that beautiful name, clings to the future. The mustard is indeed the least of all seeds, but when it is grown, it is the greatest among herbs and becometh a tree; so that the birds of the air come and lodge in the branches thereof.

Religion therefore, as I understand it, is no formula of confession, it is a moral act, it is the soaring above the lower life of animal nature. And religious faith is not a belief in something that has happened two thousand years ago: it is neither the acceptance nor rejection of the story of David's son born of a virgin, the pathetical story of the heroic martyr who died at the cross and is believed to have risen from the grave bodily. Religious faith is the confidence that we can do our duty, that we can gain the victory of spirit over matter, and that we can achieve the conquest of death.

It is death that makes it necessary for man to have religion. If there were no death in the world, we would not be in need of religion. But death, the stern messenger of eternal peace, awaits every one of us. If death did not exist, we might as well think that man is born to live happily and enjoy as much as possible the pleasures of life. But there is the pale phantom that hovers over us day and night. We know not when it will call us to the silent rest in the grave, but we do know that it will call and take us away from the circle of our family and friends, away from the field of our activity and labors.

There are some men who live like animals from day to day without giving a thought to death and without care of what may come after them. That is no life worthy of a human being. They do not fear death, it is true, but not because they have conquered death. Like the brute they do not fear it—like dumb cattle that are driven to the shambles without knowledge, without a consideration of their fate.

Life is a serious duty; and the experiences of life should teach us to number our days, that we may apply our hearts unto wisdom.

If you ask me what Religion is, I say: Religion is the creation of a higher life and the laying up of imperishable treasures. Religion is the conquest of death.

THE PRICE OF ETERNAL YOUTH.

An unnecessary dread of death prevails among mankind, a dread which is due only to a morbid imagination. Men who are not afraid to suffer pain, are sometimes found to shrink from the mere idea of hazarding their lives. It is not the agonies of death of which they are afraid, nor is it the state after death, the eternal rest of being dead, which appears appalling, but it is the moment of dying,—that it is which they dread most. It is the passage from life to death, the passage through that gate,

> "Which every man would fain go slinking by—
> Where fancy doth herself to self-born pangs compel,
> Around whose narrow mouth flame all the fires of hell."

This dread is unnecessary; it is founded upon wrong ideas of death; it is based on errors that can and must be dispelled.

We learned in school that the old physicists believed in a *horror vacui* and explained from it certain natural processes. This *horror vacui*, as we now know, is an error, just as much as the dread of death.

It is a fact that dying persons are, as a rule, under the impression that they have passed through a crisis for improvement, for the agony is overcome and pain has ceased. The feeling is due to a blunting of our sensory nerves and organs, and must be compared to the pleasant sensation which a fatigued person enjoys when quietly falling asleep. In sleep the sense-impressions become gradually dulled and sweet visions of dreams rise before our mental eye, until the light slumber passes into a profound sleep where all consciousness ceases. There is no more reason for the dread of death than for horror at lying down to sleep.

A sage of antiquity said: "Why should we fear death? Death is not here, so long as we are here. And if death is here, we are no longer."

We must meet death in the sense that the Stoic philosopher on the throne prepared himself to accept all the gifts of nature. He said: "Everything harmonizes with me which is harmonious to thee, O Cosmos. Nothing for me is too early nor too late which is in due time for thee. Everything is fruit to me which thy seasons bring, O Nature. From thee are all things, in thee are all things, to thee all things return."

Death is a natural phenomenon not less than birth; and the agonies of death are generally less painful than the throes of birth. The problem of death is closely interwoven with the problem of birth, so that you cannot disentangle the one without unraveling the other.

Birth is, as our scientists teach, the growth of an individual beyond its individuality. It is the nature of living beings to live and to grow. The lowest kind of animals do not die; they grow and divide and thus they multiply. The amœba may die from violence, it can be crushed to death by your foot; it may starve from lack of food; but it knows no natural death. The animalcules which you can observe to-day are the very same creatures that lived millenniums ago, long before man appeared upon earth. Immortality is their natural state.

How did it happen that death came into the world

of life, into the realm of immortality? Is death the meed of sin that is due to a violation of nature's laws? Or if it is a natural process, pray what is it?

Death came into the world as the brother of birth, and death became necessary when birth with its rejuvenescent power lifted organic life one step higher in its evolutionary career, so as to allow a constantly renewed progress, so as to create innumerable fresh beginnings and to give new starts to life, new possibilities to the development of life.

Birth is growth beyond the limit of individuality. Thus the creature born is the very same creature as its mother and its father, just as much as the two amœbas are the very same substance the mother amœba was before her division. But the creature born has one great advantage over its parents. It can commence life over again. It is identical with its parents, but it is its parents in a state so little fixed and formed, so young, so unimpaired, so pure, like the fresh dew that glitters in the morning-sun, that it can make a new start, it can travel new paths and can climb to higher planes, which seemed inaccessible to its ancestors.

Not only men but all creatures are naturally one-sided; they develop to be one-sided through their occupations and their experiences, and become more and more so the longer they live. What can life wish for better, than to be allowed to drop again and again the fresh prejudices constantly acquired, which we even admit may be justified in the men that hold them. But we know that they would become injurious if mankind clung to them forever. It is for the best of humanity, that it can drop the errors which are perhaps, as we freely grant, partial truths. Humanity must gain not only renewed vigor, but also virginity in life and in love, in hopes, and in ideal aspirations.

This is done through the introduction of birth into the empire of life. And this makes it possible that life is always young, that it is virgin-like, and endowed with renewed courage as well as interest.

Is the boon of a constant rejuvenescence of the race through birth bought too dearly by the surrender of our individual existence to death? Certainly not, if the good features of individuals can be transmitted to their descendants, if their death is only a partial obliteration of life, where it has lost the capacity of progressive endeavor, where impartiality of judgment is gone, so that we no longer can see the light when a new morn dawns with greater and higher possibilities.

Nature does not intend to ossify life, it makes life plastic, and in order to preserve the plasticity, the vigor, and virginity of life, nature endowed life not only with immortality that through the act of birth makes life extend and grow beyond the limit of individual existence, but at the same time it bestowed upon it, through the same means of birth, that wonderful desirable gift, eternal youth, without which immortality would become an unbearable burden.

What would life be, what would immortality mean, if it were not identical with eternal youth? If humanity must buy eternal youth at the cost of death—at the cost of the death of individuals, it is certainly not bought too dearly.

Death then is a necessity; but serious though the idea of death must make our thoughts, it is not terrible; awful though it may be, it must not overawe us. Death is like the northern sunset. The evening twilight indicates the rise of a new morn. The nocturnal darkness of the end of life is the harbinger of a new day, clothed in eternal youth. So closely interwoven is death with immortality!

The lesson that death teaches let me express in the words of our poet:

> " So live that when thy summons comes to join
> The innumerable caravan that moves
> To the pale realms of shade, where each shall take
> His chamber in the silent halls of death,
> Thou go not, like the quarry slave at night
> Scourged to his dungeon, but, sustained and soothed
> By an unfaltering trust, approach thy grave
> Like one who wraps the drapery of his couch
> About him, and lies down to pleasant dreams."

RELIGION AND IMMORTALITY.

JOHANNES SCHERR, one of the most zealous of infidels, who used all his great historical scholarship and philosophical acumen to forge fatal shafts to hurl at religion, says in one of his lucid sketches:

Religion is a groping from the Temporal into the Eternal; a pathfinding from the Finite into the Infinite; a bridge-spanning from the Sensible to the Supersensible. If we follow—and I speak now only of men who have the material and the courage to think logically—if we follow, I say, this idle worry and contention to its deepest root within us, we shall find it to mean this: terror at the thought of inevitable dissolution, abhorrence of imagined void, dread of death. Man yearns for existence beyond the bounds that are set to his life. The happy man, that he may further enjoy in a kingdom to come the comforts he possessed on earth. The unhappy one, that he may find in the land "above" the fortune he was robbed of "below." And the ideal enthusiast, that he may at last arrive at those "regions bright," where "pure forms dwell" —the prototypes of the Good, the True, and the Beautiful. Only men who through and through are men, who, in the beautiful words of Lucretius, have advanced to the point where they are able *pacata posse omnia mente tueri*, can sternly face the inexorable thought of the annihilation of the Ego and the Self, and, when the last hour is come, say with stoic resignation in the words of Manfred, "Earth, take these atoms!" The others, the millions and hundreds of millions, all wish to gain "salvation"; which means, to live beyond the grave and after death. And since it is the fashion of man to believe and to hope what he wishes, so do they believe and hope that their dear Self is "immortal" and predestined, after corporeal death, to be promoted to a higher class in the eternal school of perfection, or, as the pious in current parlance term it, "to behold God."

Scherr characterizes religion very well as the dread of death, and as a desire to live beyond death. And truly, he is right when declaring that with many religion is nothing more than the desire to make their dear ego immortal. But Scherr is decidedly wrong when he looks upon death as a finality. It is not matter alone that man consists of, but his form also; and his humanity lies not in the clay but in the spirit. In order to sustain animal life, it is sufficient to eat and to drink; but to sustain spiritual life, man must be nourished with thoughts. Our children imbibe their mental existence from parents and instructors, and the ideas with which they are reared are the very souls of the heroes of past ages; they are the souls of their ancestors and the valuable results of the lives of the departed.

The earth takes part of its atoms again in every moment of life, and it is not the atoms that we must care for most. Man does not live by bread alone, the nourishment of his soul is the word; and the word makes of him a human being. Man's life is not ended when all the atoms that shape his body return to the dust from which they came. Nature has devised means to preserve that which is human and to let the soul of man continue even after death.

I read of late in an historical essay some sentences to the following purport: 'American freedom was not possible but for the determination and strength of the Puritan character. The Puritans were not possible but for Luther, and Luther was not possible but for Paul.' If that is so, and I expect there is no one who will dispute it, can it be said that death was a finality to Luther or to Paul? When the earth took the atoms of these men, did the earth really take their whole being? No, it did not. Their better parts, those elements of their souls which were pure and noble, were preserved and will be preserved as long as men live upon earth. The ideals which they aspired to, the truths which they taught, are immortal. And like the torch in the mysteries of Eleusis that passed from hand to hand, their soul-life will be handed down faithfully from generation to generation.

The purpose of religion, indeed, is the preservation of the soul. The preservation of the soul beyond death is no illusion, no chimera of fanatic minds. It is a fact of our experience, it is a reality that can be scientifically proved. Death is no mere dissolution into all-existence. Certain features of our soul-life are preserved in their individuality. Copernicus still lives in Kepler, and Kepler in Newton; and to-day Copernicus lives in every one of us who has freed himself from the error of a geocentric conception of the world. The progress of humanity is nothing but an accumulation of the most precious treasures we have—it is the hoarding up of human souls.

SPIRITISM AND IMMORTALITY.

SPIRITISM must be well distinguished from *Spiritualism*, although in popular speech the latter term is generally employed for the former. Spiritualism is that philosophical view which, in opposition to materialism, assumes spirit as the ultimate and universal principle from which the phenomena of the world are to be explained. Spiritism is the belief in spirits and the apparition of spirits. While spiritualism is a lofty conception of profound thinkers (such as Berkeley and Fichte) who boldly spiritualize the whole universe, spiritism, on the contrary, materializes even spirit itself and spiritual phenomena. With the spiritist the spiritual realm has become a world of spirits.

Spiritism, and the belief in spirits, may often have been occasioned either by successful impositions, or by mysterious phenomena, which for a length of time frustrated all attempts at explanation. But, ultimately, the origin of spiritism lies deeper; the source from which it is nourished, is man's longing for immortality. From the vague hope of life beyond the grave, and from the dread of being entirely annihilated, spiritism draws its strength; and all attempts at disclosing the deceptions of impostors, and at explaining certain marvelous phenomena which had been regarded as certain proofs by believers, will remain futile, so long as the spirit of man is considered as an entity, existing of itself and inhabiting the body during the time of life. The idea of immortality which is an exceedingly powerful factor in human emotions, must, in this combination, produce the most fantastical and nondescript errors, which, wherever they have been implanted, will take firm root in the human mind. Not that the errors possess that strength of themselves; they derive it from the truth with which they are mingled; and a total annihilation of ourselves is so utterly inconceivable, that we feel by an instinctive intuition, as it were, the truth of immortality.

The immortality of the soul is commonly understood to be the continuance of our conscious ego beyond death in the shape of a spiritual, bodiless being. This view rests on the principle that the soul is an entity which inhabits the body and can exist of itself; accordingly the ego is considered as a substance which is supposed to be the constant and continuous factor behind the transient states of consciousness.

The immortality of the ego stands and falls with the belief in a ghost-soul, and the only scientific evidence for the existence of a ghost-soul has been the supposed unity of consciousness. If our consciousness were a substance, and if, as a substance, it possessed a unity, for instance like that of an atom, the ego of our consciousness would perhaps be indestructible. Kant's Critique of Pure Reason teaches that the ego, as an entity, is a fiction. We are aware of a series of ideas that become conscious in our mind. It is these ideas that are constantly present, but to consider consciousness as a substance that exists apart from its contents of ideas is an illusion, a fallacy or paralogism of pure reason. Modern investigations in physiological psychology show that the ego, with its chains of conscious and subconscious states, is the product of many factors under very complicated conditions. The ego forms a unity, *i. e.*, a unitary complex, or a compound system; but, of itself, it is not a unit. The *Einheitlichkeit* of the soul must not be construed as a rigid and ultimate *Einheit*. In a similar way the French school of experimental psychology, foremost among them Th. Ribot and Alfred Binet, have proved that the ego is not an entity constituting the "cause" of mental phenomena, but, on the contrary, that the ego is the "effect" of certain phenomena of mental activity. If this ego, as an entity, is an illusion, how can it be immortal? If a ghost-soul does not exist, how can it continue to exist? If a consciousness independent of its contents, which are the ideas that become conscious in our mind, has no reality, how can we attribute to it a permanence *in æternum?*

Although a ghost-immortality of disembodied spirits is impossible, man's existence is not a fleeting phenomenon of an ephemeral nature. His soul-life is not of yesterday, and does not vanish into nothingness to-morrow. His ideas as well as his actions are facts that continue to be factors in the future development of his race. The life of a single individual is not a separate and single event that begins with his birth and disappears again at his death. It is the product of a long evolution of many thousands of generations. Their works and thoughts live in the present generation, and our soul-life, our thoughts, accompanied with the same kind of feelings, will continue to exist in the future. Those who think, who act, and who feel like ourselves, possess our souls,* and *in them we* shall continue to live and move and have our being.

* Compare THE OPEN COURT, page 396, first column, lines 1—11.

It is objected that, as a rule, people do not care for such an immortality; they want the immortality of a ghost-soul. This is undoubtedly true, but whether they care or not, it does not alter the facts. If people do not care for this grander kind of immortality, they must be educated to appreciate it.

A Christian missionary in Greenland told his Esquimaux converts much about their future life in heaven, and when he was asked whether there would be plenty of whales and seals and walruses, and whether the redeemed would have enough cod-liver oil, he suggested that they would no longer want such things. The Esquimaux then turned away and said: " What is the use of your heaven if there are no whales, nor seals, nor walruses, and if we can have no cod-liver oil. If such things don't exist, and if the most glorious joys are not even desirable in heaven, we don't care for it at all."

Similarly among us, those people who believe that the soul is a ghost which inhabits the body, do not care for any immortality unless it be that of a ghost-soul. They do not care for continuing to live in the life of mankind, and are satisfied to hover about as spirits, communicating with their beloved ones through raps and other primitive manifestations. They are like the prodigal son, who left his father's house and fed upon husks for want of better food.

All the most marvelous feats of mediums do not attain to that wonderful perfection for which our best performers in legerdemain are famous. The ingenious way in which they present their clever deceits is also truly remarkable. The worst thing about spiritism is its dearth of ideas. The spirits show in their communications an extraordinary lack of spirit. If the manifestations were as true and undeniable as daylight, they would reveal a most pitiable state of spirit-life, "sans teeth, sans eyes, sans taste, sans—everything."

It is impossible to convince a spiritist of his errors simply by showing that he has allowed himself to be duped—so long as he believes in the immortality of a ghost-soul. The idea of immortality is strongly implanted in the human mind, because every living being feels that life cannot be annihilated; as Goethe says:

> "*Kein Wesen kann zu nichts zerfallen,*
> *Das Ew'ge regt sich fort in allen,*
> *Am Sein erhalte Dich beglückt!*
> *Das Sein ist ewig; denn Gesetze*
> *Bewahren die lebend'gen Schätze*
> *Aus welchen sich das All geschmückt.*"

> " No being into naught can fall,
> The eternal liveth in them all.
> In All-Existence take delight,—because
> Existence is eternal; and fixed laws
> Preserve the ever living treasures
> Which thrill the All in glorious measures."

This consciousness of our indestructibility is so direct and immediate that, in a healthy state of existence, we feel an eternity of life in every moment, and only with the assistance of much contemplative thought and earnest reflection can we conceive at all the idea of death. Even if this earth, the intellectual life of which has found its consummation in mankind, should break to pieces and make a further and direct continuance of our ideas, our actions, and our soul-life impossible, we know that new life will grow from the wrecks of our world; that new suns will shine upon new planets peopled with new generations, who, like ourselves, will aspire to the same aims and struggle for similar, perhaps even higher, ideals.

The idea of immortality resting on a true instinct, and on the natural conviction of the indestructibility of life, cannot be easily blotted out from the human mind, even though mixed with errors. And the idea of immortality need not be eradicated; we have simply to weed out the errors that grow around it by the slow and long process of patient education. Those who have freed themselves of the old errors that have attached to the conception of immortality look smilingly upon their former views, as the man thinks of his having been a child with childlike thoughts. As the Apostle says: "When that which is perfect is come, then that which is in part shall be done away. When I was a child, I spake as a child, I understood as a child, I thought as a child; but when I became a man, I put away with childish things."

The old view of considering our ego as a real entity is, as the sacred Hindoo religion expresses it the veil of Maya that lies upon our eyes. The man who recognizes this ego to be a sham has become a Buddha, *i. e.*, a knower—one who knows; one from whose eyes the veil of Maya has been taken. He no longer lives the sham-life of egotistic desires that moves in the circle of never satisfied wants, but he has entered Nirvana. The annihilation of the ego is the condition of a better life, of a broader and higher existence.

This truth, though not fully realized in Buddhism, was nevertheless presaged by its great founder, Gautama. It has been mixed with pessimistic vagaries and monstrosities, but has at the same time afforded comfort to millions of people in their troubles and cares and agonies of death. This same truth is the basis of the Christian religion also, whose founder demands a surrender of our egotistic desires. Christ says: "Whosoever shall loose his life shall preserve it."* And this same truth lies at the bottom of all true ethics. We must entirely surrender our ego and regulate all our actions by a maxim fit to become a universal law (as Kant expresses it). By lifting all our thoughts and intents to the broader interests of promoting life and of promoting higher forms of life, we cease to be single and separate beings, and become the representations of cosmic life, or in biblical terms, "The householders of God."

The surrender of the ego is a destruction of self and of selfishness only, but it does not imply, as has been assumed by pessimistic teachers and by the monks of a world-despising attitude, an annihilation of our existence and of life generally. It does not mean death, but life; not inactivity, but work; not destruction, but immortality. It means life and progress and aspiring labor, not in the service of egotistic purposes, but for the evolution of existence in higher forms, for the development of our race and the realization of the ethical ideal.

All labor for egotistic purposes would be vain, for, we shall die, and the purpose for which we have worked would be gone. But if we consider ourselves as householders who stand in the services of a higher purpose than ourselves, if we aspire for a further evolution of cosmic life: the purpose of our lives will not die with us; we shall continue to live in our deeds and

---

* The same idea is almost literally (though with the addition of "for my sake") repeated over and over again. Luke xvii. 33; Luke ix. 24; Matt. x. 39; Matt. xvi. 25; Mark. viii. 35; John xii. 25; John x. 17.

thoughts and in those who are inspired by the same ideals; as Schiller says:

> "Art thou afraid of death? Thou wishest for being immortal?
> Live as a part of the whole; when thou art gone it remains."

This view of immortality is not less, not smaller and more meager, than the immortality of a ghost-soul, whose very existence is an unwarranted assumption. It is more; it is grander and sublimer; although those who have the veil of Maya upon their eyes, who still believe in that sham-entity of the ego, cannot understand and appreciate it.

Johannes Tauler, of Strassburg, one of the profound mystic preachers of the beginning of the fourteenth century, said: "*Wir müssen entwerden, um Gott zu werden.*"* Our ego must be undone in order for us to become God. The higher life of immortality will be ours; but the price to be paid for it, is a surrender of the sham-existence of our ego.

* Quoted from memory.

IMMORTALITY AND SCIENCE.

It appears as though the problem of immortality had to be solved anew by every generation. How often has the question "When a man dies shall he live again?" been answered in the affirmative as well as in the negative? But it appears that a final answer has not as yet been given. Before the court of science the religious answer "Man *shall* live again!" is a mere assertion. It is the expression of a sentiment, and we may grant that the sentiment is quite legitimate, it is a strong sentiment, and to many people it is the most religious, the most sacred sentiment. It is a holy hope without which they cannot live. How deep the roots of this sentiment are buried in many souls will be seen from the following extract from a letter which I received from a well educated gentleman whose life has been spent in teaching and who was devoting the leisure of his old age to philosophical studies. Having explained some of his scientific doubts concerning the immortality of the soul and having rejected at the same time the arguments that are generally brought forth against this belief, he adds these thrilling words:

"I am now seventy-four years old, but instead of growing more cheerful and assured, the reverse has been the case. Accordingly my present state of soul is lamentable and pitiful. Whether

I shall end my life in distraction and insanity or in confidence in myself and God, I cannot say."

Granted that the belief in immortality is a legitimate sentiment; it may be a postulate and an indispensable condition of our religious life, yet as long as it remains the mere expression of a sentiment, it is one-sided and insufficient.

However, the unbeliever's answer, which so often boasts of being the voice of science, is no less one-sided. And the denial of immortality is religiously not so heterodox as most unbelievers suppose, for it has been forestalled in the Biblical sentence of Solomon:

> "I said in my heart concerning the estate of the sons of men that God might manifest them,* that they might see that they themselves are beasts. For that which befalleth the sons of men befalleth beasts; even one thing befalleth them; as the one dieth, so dieth the other; yea they have all one breath; so that a man hath no pre-eminence above a beast: for all is vanity. All go unto one place; all are of the dust, and all turn unto dust again." Solomon in Eccl, 3, 18-20.

It appears from this quotation that either side of the question is quite biblical.

* * *

Goethe says:

> "'Hast immortality in mind
> Wilt thou thy reasons give?'
> —The most important reason is
> We can't without it live."

The belief in immortality is of paramount importance because it is a moral motive. It is perhaps the most powerful moral motive man has, and it is of great importance because if man regulates his life as if he

---

* The Hebrew *leharam ha Elohim* is more correctly translated in the Septuaginta, ὅτι διακρινεῖ αὐτοὺς ὁ θεός "that God will distinguish them." The sense is: I pondered on the nature of men, whether God distinguishes them, but it appears that they are beasts.

were immortal, he will survey a larger field than if he limits his interests to the narrow span of his own individual life. In other words, the belief in immortality is useful; it induces men to adapt themselves more fully to the great social organism of mankind; it makes their life more moral. On this account it has been proposed: Let us foster the belief in immortality among the masses, although it may be untenable as a scientific conception.

This proposition has been called a *pia fraus*—a name invented for its justification, and the pious fraud method has sometimes received more credit than it deserves. Is it necessary to add that pious fraud should be denounced as immoral and objectionable under all circumstances?

If, however, the belief in immortality is indeed useful, I maintain that it must contain a truth. A falsity may be useful once or twice, or a hundred times, but it cannot be useful in the long run, for centuries and millenniums. The belief that death is no finality and that man shall live again, which so generally prevails in all our many churches and religious societies contains a truth in spite of the apparent and undeniable counter-truth that man is "like grass which groweth up; in the morning it flourisheth, and groweth up; in the evening it is cut down and withereth." [Psalm xc, 5-6.]

What is this truth? Has science, especially through the discovery of its latest great truth, the doctrine of evolution, shed any new light upon the problem? and if it has, what is the new conception of immortality as it appears from the standpoint of the evolutionist?

The question of immortality is not beyond the pale of science. It is not only our right to investigate whether man's instinctive longing for a continued existence is justified, it is also our duty to attain to clearness concerning one of the most important and basic problems of psychology, and also of ethics. Also of ethics! For the immortality idea forms the centre of all ethical questions. It affords the strongest motive to moral action. Indeed what is morality else but the regulation of our actions with an outlook beyond the grave, it is a building up not only sufficient to hold for our life-time, but for eternity.

* * *

All living beings have a dread of annihilation; everything that exists has a tendency to continue its existence; and it will continue to exist, for there is no annihilation. Being can never change into not-being. There is annihilation only in the sense of dissolution. A certain combination ceases to exist in this form because it changes into other forms. Being exists, it is eternal, and it cannot be annihilated. Not-being does not exist and will never exist. Not-being is a non-entity, a mere fancy of our imagination. There is no reason whatever for anything that exists to fear annihilation. We may dread change, but we need not dread annihilation,

Our dread of losing consciousness is not justified. We lose consciousness every night in sleep, and it is a most beneficial recreation to us. The boiling water may be afraid of being changed into vapor. But its fear is groundless; nature will again change the vapor into drops of water. From the surface of our planet all organised life may die off. Our solar system may crumble away into world dust, but what is that in the immeasurable whirl of suns? There are other parts of the milky way in which new worlds are forming themselves, and we have sufficient reasons to believe that the tide of life ebbs and swells in the whole universe not otherwise than autumn and spring change alternately in the northern and southern hemisphere on this planet of ours, not otherwise than waking and sleeping, activity and rest, day and night change in our lives. The single forms of life can be destroyed, but life remains eternal; life is indestructible, it is immortal.

This truth has been maintained again and again; et many declare that it gives no satisfaction to them unless their persons are included in the general law of preservation and it is generally supposed that before the tribunal of science there seems to be little chance for proving the persistence of personality.

Nevertheless, there is a truth even in the idea of the preservation of the individual soul, and we do not hesitate to say that it is the most important aspect of the immortality idea. That the individual features of our souls are preserved has been proved by evolution. Evolution takes a higher view of life. It considers the whole race as one and recognises the continuity of life in the different generations.

Humanity lives and the individual is humanity incorporated in a distinct and special form. Humanity continues to live in spite of the bodily deaths of the individuals—and truly it continues to live in the distinct and special, in the personal and most individual forms of the individuals. Bodies pass away, but their forms are preserved and their souls are here still. The preservation of experience from generation to generation, is the condition of intellectual growth. The preservation of that which is contained in and constitutes the very personality of man is the basis of progress. In one word the immortality of the soul makes its higher evolution possible.

Evolution teaches a new conception of the soul. It destroys the old-fashioned idea of an individual. It shows that the birth of an individual so called is not a new beginning, but it is only a new start of prior life. The baby which is born to-day is a product of the sum total of the activity of its ancestors from the moment organised life first appeared upon earth. And organised life, what else is it but a special form of the cosmic life that animates the whole universe?

What is man's soul but his perceptions and thoughts, his desires, his aspirations and his impulses which under certain circumstances make him act in a certain way. In short, man's soul is the organised totality of his ideas and ideals. These ideas and ideals of man have been formed in his brain through experience which is transmitted from generation to generation, and in preserving them we preserve the human soul.

Man's soul is not the matter of which he consists at a certain moment. Man's soul is that particular activity of his which we call his thoughts and motives. So far as our brother has the same thoughts and the same motives, he has also the same soul; and since the doctrine of evolution has become a truth recognised by science, we can with a deeper meaning repeat the ancient saw of the Hindoo sages, "*Tat twam asi—* That art thou." All living creatures are ourselves; they are in possession of souls like ourselves, and the more they feel and think and act like ourselves, the more have they our souls.

It is true that from this standpoint our souls are not something exclusively our own, they are not, as it were our private property. Our souls are in part inherited and in part implanted into us by education. The former part consists chiefly of our physical constitution and general disposition, the latter part embracing our thoughts and ideals is by far the most important one; it represents the highest and most human elements of our souls.

There is accordingly a truth in the Buddhistic doctrine of a pre-existence and migration of souls. And this truth holds good for the past as well as for the future. Soul is not an essence, but a certain kind of activity; it is a certain form of impulses, on the one hand conditioned by innumerable experiences of the past—"inherited memory" it has been called by physiologists—and on the other hand conditioning in its turn the future. This latter fact, viz. that our present soul-life is conditioning the future, it will at once be understood, is the most important ethical truth. It must be borne in mind when we are about to act, that every act of ours continues in its consequences. The act may be unimportant, and the consequences may be unimportant too, nevertheless it continues with the same necessity as that every cause has its effect.

Death is no finality, and we must not form our rules of conduct to accord with the idea that the exit of our individual life is the end of all. Says W. K. Clifford in his article "The Unseen Universe":

> "The soldier who rushes on death does not know it as extinction; in thought he lives and marches on with the army, and leaves with it his corpse upon the battle field. The martyr cannot think of his own end because he lives in the truth he has proclaimed; with it and with mankind he grows into greatness through ever new victories over falsehood and wrong.
>
> For you, noble and great ones, who have loved and labored yourselves not for yourselves but for the universal folk, in your time

not for your time only but for the coming generations, for you there shall be life as broad and far-reaching as your love, for you life-giving action to the utmost reach of the great wave whose crest you sometimes were!

The preservation of the special and most individual contents of man's personality, the preservation of that something in him which he regards as the best and most valuable part of him is the strongest motive for moral action. Even an unclear idea of the immortality of the soul is therefore better and truer than the flat denial of it. And this is the main reason why the churches survived in the struggle for existence against those people who looked upon death as an absolute finality. The ethics and ethical motives of the churches come nearer the truth than the ethics of those who believe that the death of the individual ends all of the individual, body and soul.

* * *

Here I might rest my case. But I feel that those who attach to the belief in immortality the idea of a transcendent existence in some kind of heaven, are disappointed because I have not as they suppose, touched the most vital point of the subject. I grant that from their standpoint, I am guilty of this mistake. The reason is that I have tried to state the positive view of the problem and not its negative aspect.

Immortality means the continuance of life after death; continuance means a further duration of the present state. If you mean by immortality, the soul's existence in the shape of a bodiless ghost, you should first prove the existence of bodiless ghosts. Our experience knows only of souls which are the activity of organisms in their awareness of self. You cannot preserve what you do not have, and you should not worry about losing something you never possessed; in fact you cannot lose it. If immortality of the soul means an existence as pure spirit, this would not be a continuance of life after death, but the new creation of an entirely different being about the mere possibility of whose existence we can form no more a conception than about an immaterial world in which there would be no display of forces. What is the use of racking our brains as to whether an ethereal world can exist and what comfort can we derive from a belief in its possibility?

The old view of "the resurrection of the body" as it has been worded in the apostolic creed, is certainly more in agreement with modern science and with the doctrine of evolution, than the later belief of a purely spiritual immortality.

* * *

Let me add here a few words in answer to the anxiety of the old philosopher who finds himself on the verge of despair because his hope in an unbroken continuance of his consciousness after death somewhere in an unknown cloudland finds little or no support in science. The scientist, the philosopher, the thinker, should never trouble himself about the results to which his inquiries lead. A sentimental man who wants his preconceived views proved, who hopes for a verification of favorite ideas, is not fit to be a thinker. I do not mean to say that sentiment is not right, but that sentimentality is wrong. It is not right that sentiment should perform the function of thinking. Thinking requires courage and faith, it requires faith in truth.

Truth often appears to destroy our ideals. But whenever it does destroy an ideal, it replaces it by something greater and better. So certain features of the old immortality idea are untenable before the tribunal of science; yet the idea of immortality which is taught by science, is surely not less sublime, not less grand and elevating than the old one. It teaches us not only a general persistence of all that exists, but a continuance even of that which constitutes our personal individual life.

In looking around and studying the facts of life, we find that we can everywhere improve the state of things; there is no place in the world where there is no chance for improvement, for useful work, for progress. Yet there is no chance whatever for improving the cosmical conditions of the world, the order of the universe, or the laws of nature. And truly it is good for man that he cannot interfere here, because he could never succeed with his improvements. Dominion is given to man over the whole creation, but his dominion ceases where the divinity of nature, the unchangeable, the eternal, the unalterable, of cosmic existence begins.

If there is a God, it is this something "that is as it is," expressed by Moses in the word "JAHVEH." Confidence in God, if it means that we expect *him* to attend to that which can be done by *ourselves* is highly immoral, but confidence in God in the sense that the unalterable laws of nature just as they are, are best for us and for everything that exists, and that it would be mere folly on our part to wish them to be different, is a great truth, and belief in it is no superstition; it is true religion, it is the faith of the scientist, of the philosopher, of the thinker; it is our trust in truth.

The idea of a purely spiritual, a transcendent immortality would be possible only if the name and being of Jahveh, if the revelation of God in the reality of nature were either a great sham, a lie on his part, or a huge error on our part. The view that nature is unreal and that outside of this great cosmos of ours exists another and purely spiritual world is called dualism. There are no facts in experience to support dualism or a dualistic immortality. However, the idea of an immanent immortality is based upon facts demonstrable by science. It is an undeniable truth—undeniable even by the dualist, who in addition to it believes in a purely spiritual immortality somewhere beyond the skies.

Goethe whose view of life was an harmonious and consistent monism, expresses his belief in immortality in the following lines:

> " No being into naught can fall,
> The eternal liveth in them all.
> In all-existence take delight—because
> Existence is eternal; and fixed laws
> Preserve the ever living treasures
> Which thrill the All in glorious measures."

DEATH, LOVE, IMMORTALITY.

How is it that our poets so often set into opposition the ideas, love and death? Is there a secret connection between them? and if so, can that connection be explained?

The Hebrew poet in the song of songs, sings:

> "Set me as a seal upon thine heart,
> As a seal upon thine arm:
> For love is strong as death;
> Jealousy is cruel as the grave;
> The coals thereof are coals of fire,
> Which hath a most vehement flame.
>
> "Many waters cannot quench love,
> Neither can the floods drown it;
> If a man would give all the substance of his house for love
> It would utterly be contemned."

Love is strong as death, nay it is stronger; for if there is any power that can conquer the grimmest foe of man, it is love. Love therefore, as the conqueror of death, represents immortality.

How many foolish conceptions of immortality obtain among mortals, and how often have they been refuted by the sages of all creeds and of all philosophies! Nevertheless, the belief in immortality is as firmly rooted in the souls of men to-day as it ever has been in past ages. We have of late read that beautiful passage of the American heretic who rejects all religion, who hates Christianity, and is in every respect an unbeliever. He has no ridicule, no flippant word however, for immortality; he says:

"The idea of immortality, that, like a sea, has ebbed and flowed in the human heart, with its countless waves of hope and fear beating against the shores and rocks of time and fate, was not born of any book, nor of any creed, nor of any religion. It was born of human affection, and it will continue to ebb and flow beneath the mists and clouds of doubts and darkness as long as love kisses the lips of death.

"I have said a thousand times, and I say again, that we do not know, we cannot say, whether death is a wall or a door—the beginning, or end, of a day,—the spreading of pinions to soar, or the folding forever of wings—the rise or the set of a sun, or an endless life, that brings rapture and love to every one."

What is death? Is it not the destruction of that form of ours after it has become unfit for further use? It is maintained by the agnostic orator that we cannot know whether it is the rise or the set of a sun. Let me answer that to us death appears like the set of a sun; but we know that the sun itself never sets.* As its light never ceases to shine, so life is immortal.

What is love but our longing for immortality? And the old man who looks upon his youthful sons and enjoys the baby-smiles of his grandchild,—does not a new vista of life open to him? And is not that life that beams in the eyes of his children and grandchildren his very own life? Does he not commence a new career in every one of them? Is it mere sentimentality, an empty figure of speech if we say that love has conquered death indeed? Let death have its prey, if we but live again, if instead of remaining as we are, small, limited, egotistic, we may grow and expand, if new chances of commencing life over again are given unto us, and if guided by love we can determine ourselves,

---

*Cf. Schopenhauer, Parerga und Paralipomena, über die Unzerstörbarkeit unseres Wesens durch den Tod.

how we may be improved in future generations! Let death have its prey, if our better selves, our noblest thoughts, our highest ideals, our best deeds will live in, and have a beneficial effect upon, future generations.

Love is not limited to sexual love. Love is enthusiasm for everything good and great; love is every true and noble idea worth being thought again and again, and to be propagated to the most distant generations.

Our body, the visible appearance of our ego, is sure to die; and there is no ground for bewailing it, for what is the use of preserving just this combination of dust with all its little defects,—a combination whose psychical components are a medley of a few true ideas, of a few lofty aspirations mixed with errors and prejudices? Is it worth while to preserve this alloy as it is? O no! It is a thousand times more preferable to preserve the good, the true, the ideal thoughts only, as Nature really does, and let errors as well as prejudices perish as they deserve.

Immortality is no fiction, and a craving for immortality is a natural feeling of the human heart. True immortality is not possible by egotism, for there exists no such a thing as an immortality of the ego. True immortality is realised by love only; and love is not only the affection toward our beloved ones; love is every aspiration for truth, every hope for progress, and every enthusiasm for the ideal. Love is the broadening of our soul beyond the limit of the ego. But it is not enlarged egoism either; love has always something of a humanitarian and a universal spirit. It thrills our pulses with the life of the All and grants in a fleeting moment the bliss of a whole eternity.

Immortality is not presented to us by some generous donor as a gift. It must be gained by our own efforts; by our struggling for it must it be deserved. But there is that comfort in it that it can be gained by every one who believes in Love.

In this spirit the German poet says:

> "Out of life there are two roads for every one open :
> To the Ideal the one, th' other will lead unto death.
> Try to escape in freedom as long as you live, on the former,
> Ere on the latter you are doomed to destruction and death."

FREETHOUGHT, ITS TRUTH AND ITS ERROR.

By freethought we understand the right of every thinker to seek for, to find, and to state the truth himself, and in calling freethought "a right" we are well aware of the fact that as all rights are only the reverse of duties, so freethought is at the same time the duty of every thinking being to seek for, to find, and to state the truth for himself. And this duty, in our conception of religion, is also the highest religious duty of man. The religion of science, therefore, may also be called, in this sense, the religion of freethought.

Freethought stands in opposition to authoritative belief. There have been and there are still religious teachers and institutions which maintain that man should not seek the truth for himself, because he is, as is claimed, unable to find it, and if a man has become convinced that he has found some truth for himself, he must be mistaken and therefore he should not be allowed to pronounce it, his errors being injurious to his fellowmen.

Man accordingly, because he cannot know, should believe, he should trust in what he is told to be the truth, he should give himself and his reasoning up to the higher authority of the church, "bringing into captivity every thought" (2 Cor. x, 5). Freethought has risen in revolution to the religion of blind obedience, and freethought, although first suppressed by ecclesiastical and secular authorities, has come out victorious in the end and is now almost generally recognised as the cornerstone of progress among all the nations which represent civilised humanity.

Freethought has often been misunderstood. It is not only misinterpreted by the adversaries of freethought, but not unfrequently also by those who call themselves freethinkers. Freethought does not mean that thought is free or should be free, it simply claims freedom for the thinker to think undisturbedly and uninterfered with for himself. The thought of the thinker however is not free and cannot be free, in the sense that the thinker can think however he pleases. Freethought, it is true, claims the liberty and the right to think for the individual; but that right being procured, the individual can think only by renouncing its individuality. We can dream as we please, we can imagine that this or that might be so or so just as we like. But when we think, we cannot come to a conclusion just as we please, we have radically and entirely to give up our likes and dislikes in order to arrive at what can objectively be proved to be the truth.

The freethinker who claims not only liberty for thought, but also liberty of thought is gravely mistaken. There is no liberty of thought. The mere idea "liberty of thought" is a contradiction, for thought is strict obedience to the laws of thought and only by strict obedience can we arrive at the truth which is always the purpose and final aim of thought.

The error that there can be liberty of thought has led to another erroneous idea which is a misinterpretation of the principle of tolerance. We certainly believe in tolerance, but tolerance means the recognition of other people's right to express their opinion. It does not mean that any and every opinion is of equal value. Tolerance demands that the opinions of those who seek the truth should be heard; they should not be put down with violence or treated with contempt. Yet tolerance does not exclude criticism; it does not and should not abolish the struggle for truth among those who believe that they have found the truth. For truth is objective and there is but one truth. If tolerance is based upon the idea that truth is merely subjective, that something may be true to me which is not true to you, and that therefore an objective conception of the truth is an impossibility, tolerance has to be denounced as a superstition. Tolerance in this sense is injurious to progress, for it prevents the search for truth and leads to the stagnancy of indolent indifferentism.

The expression objectivity of truth must not be understood in the sense that truth is an object. Truth is not a thing, but a relation. Truth is the congruence of our ideas with the reality represented in these ideas. If the idea is a correct representation of the reality represented so as to form a reliable guidance in our deportment toward the reality, it is true. That truth can be more or less clear, that it can more or less be mingled with errors, that it can be more or less complete or exhaustive is a matter of course. Truth cannot be possessed as objects are possessed so that we either have it entire or not at all. Truth is the product of our exertions, it is the result of our search for truth, so that, the world of realities with its innumerable relations and unlimited changes being living before us, immeasurable, interminable, and eternal, truth can never be complete, never perfect, never absolute in the minds of mortal beings. But that proves only the greatness of the universe and the grandness of the object of our cognition. It is no fault of truth. For truth remains truth, it remains objective, and can as such serve as a guidance for conduct, even though it be incomplete and imperfect. We however are freethinkers and search boldly for a more complete and more perfect conception of truth, because we trust in truth—in its objectivity, its exclusiveness, its universality, and its authority.

Freethought, if the word is conceived as the right and the duty of everybody to think for himself, boldly abolishes the slavery of blind obedience, but it does not abolish, as is sometimes erroneously supposed, any and every authority. On the contrary, its claim is based upon authority and can be maintained only on the strength of this authority. This authority is the objectivity of truth, which involves its uniqueness. There is but one truth. All the many different truths are but so many parts or aspects of truth; and although the different aspects of truth may form contrasts, although we may state them in paradoxical formulas, they never can collide so as to enter into a real and actual contradiction. Whatever is positively contradictory to truth is impossible, for truth is one and is always in harmony with itself. Truth is objective and the right to think is based upon the confidence that correct thought which is rigidly obedient to the laws of thought, will lead to the cognition of truth.

Freethought accordingly is not the renunciation of all authority, it is only the renunciation of human authority. It is not the abdication of obedience, it is only the abdication of blind obedience. Freethought refuses to recognise special revelations not merely because it disbelieves the reports made about these special revelations, not merely because it declares them to be doubtful and unreliable. Freethought would be weak if it were based on mere negations and disbeliefs, and that freethought which never ventures farther than the negations is weak indeed. Freethought refuses to recognise special revelation, because it believes in the universal revelation of truth. The God of freethought is not a God who contradicts himself, who makes exceptions of his will by miracles for those who seek after signs. The God of freethought is not far from every one of us. We can seek him, if haply we might feel after him and find him. For in him we live and move and have our being. He appears in the realities of nature and of nature's laws, and his revelation is not dual; it is one, it is throughout consistent with itself and every one is welcome to search for the truth.

Because God has been conceived as a miracle-working magician, and because the ecclesiastical authorities have again and again maintained that such a God alone can be called a God, freethought has been driven into the negativism of atheism. But if God is conceived as the objective reality in which we live and move and have our being, as that power the cognition of which is truth and conformity to which is morality, freethought is by no means either negative or atheistic. Freethought is by no means a mere negation of belief, it is by no means an overthrow of religion, or a reversal of religious authority. Freethought is a strong and potent faith. It is the faith in truth.

The faith of freethought is as a grain of mustard seed, which indeed, is the least of all seeds, but when it is grown it is the greatest among herbs, and becometh a tree, so that the birds of the air come and lodge in the branches thereof. The faith of freethought is in the beginning a mere maxim, a hope, an ideal. But it is founded on the rock of ages; it is founded upon truth. The faith of freethought is justified. We have a right to search for the truth; yea, we have the duty to search for the truth. And why? Because truth can be cognised. Truth is not an illusion, not a mere subjective fancy, it is founded upon objective reality. It is an ideal that can be approached more and more, not a mere vision but a realisable actuality. It is a path, although a steep path full of thorns, a narrow and strait gate and few there are that find it. But we must find it for all other paths lead astray. And we can find it, and blessed are those who have found it, for it alone leads onward and upward; it alone is the way of life, it alone is the road of progress.

THE LIBERAL'S FOLLY.

There was a man in the Fatherland to whom liberty was dearer than life. He bravely stood up against the Government and against the Church, for both proved oppressive, both curtailed the liberties of the people. There was no freedom in the Old Country, and no hope of ever attaining freedom. So this man left his home and the place of his childhood; he crossed the Ocean and came to the Land where the Stars and Stripes float in the breeze as an emblem of the new ideals that have become actual facts under our western skies.

This man arrived here poor, but he was industrious, frugal, and intelligent. He worked first as a laborer, then as a mechanic, then as an inventor. He earned money and he saved money; first cents, then dollars, then hundreds, and then thousands of dollars. After a life of energetic labor he had become one of the wealthiest citizens of his adopted country.

He had children and they were educated according to his principles. They should not be suppressed, as he had been during childhood; they were brought up in liberty.

To-day this man is broken-hearted. Part of his wealth is gone, through the imprudence and folly of his son. Everybody had seen it, but the father had not, that his son brought up in liberty had become a scamp, a foolish, rude lout, a boisterous scape-grace. The father had enjoyed the pranks of the frolicking child; but he was disappointed when the adult son repeated the same pranks in business—not to mention other dissipations and follies.

Who is that man? His name is legion. Look around, and you will recognize him at every turn among your acquaintances and your business friends. This man can almost be considered as the typical Liberal. It is not always his immediate son who thus shows the folly of his errors; in many cases it is the grandson or the child of the grandson. For the virtues of the parents remain a blessing to the second and third generation. The capital of moral strength is not suddenly exhausted; yet it dwindles away rapidly.

The children of men of this stamp sometimes still remain in possession of their father's wealth. If not laborious and industrious, yet they are shrewd business men, sometimes unscrupulous too; but they have mentally and morally degenerated, and in the place of the republican simplicity of their grandsire they assume aristocratic habits. They are ashamed of the honesty, the industry, and frugality of their ancestors and make themselves ridiculous as servile imitators of European nobility.

Let us institute an aristocracy of the mind, and of loftiness of aspirations. Rotten is every nobility that boasts of wealth. It is a shame that we Americans, "the brave and the free," are always vaunting in the face of foreigners the immeasurable, inexhaustible riches of our country. It is a poor country where that is the best to be gloried about, and it is a poor man whose riches are everything of value that he possesses. Let us cease to admire the rich because they are rich ; and ye, the moneyed aristocracy, cease to pride yourself upon your possessions. The pride of wealth is the lowest kind of pride, the meanest, the poorest!

But ye liberals, beware that ye are not under the same curse as the typical liberal. Ye liberals have a great mission, for ye are the salt of the earth : but if the salt has lost his savor, wherewith shall it be salted? It is thenceforth good for nothing, but to be cast out and to be trodden under foot of men.

Liberty is a great thing and we should give, if need be, our lives for liberty. But liberty must be deserved ; it must be the fruit of our labor. Do not be deceived by the false prophets who preach in high-sounding words, who promise happiness and enjoyment, and then decoy you into the abysses of the pleasures of the world. They come to you in sheep's clothing, but inwardly they are ravening wolves ; they tell you that liberty enlightens the world. Do not be deceived, for it is just the reverse. Liberty does not bring enlightenment, but enlightenment brings liberty ; and there is no liberty which is not based on enlightment, on education, on culture, on morality, on wisdom, and good will.

The impoverished immigrant is the fool of whom the gospel speaks. His ground had brought forth plentifully, and he thought within himself, saying, What shall I do, because I have no room where to bestow my fruits? and he said, This will I do: I will pull down my barns, and build greater ; and there will I bestow all my fruits and my goods, and I will say unto my soul, Soul thou hast much goods laid up for many years ; take thine ease, eat, drink, and be merry. But God said unto him, 'Thou fool, this night thy soul shall be required of thee, then whose shall those things be, which thou hast provided? So is he that layeth up treasures for himself, and is not rich toward God. For a man's life consisteth not in the abundance of the things which he possesseth, but in the abundance and purity of his soul.

The rich man was a fool because over the cares for worldly goods he forgot the one thing that is needed. He neglected his soul; and his soul was taken from him.

The man to whom liberty was dearer than life neglected his soul and he neglected to build up the souls of his children. Thus they degenerated and involved their old father in their own ruin.

You liberals call yourselves free-thinkers and you rail from the platform at the churches and at religion. Ye blind guides! Why behold ye the mote that is in your brother's eye, but perceive not the beam that is in your own eye? Either, how can you say to your brother, Brother let me pull out the mote that is in thine eye, when you yourself behold not the beam that is in your own eye? Ye hypocrites, cast out first the beam of your own eye and then shall you see clearly to pull out the mote of your brother's eye.

How insignificant is the mote in the eye of an orthodox clergyman who when teaching morality cannot as yet dispense with the traditional fairy-tales, in comparison to the scoffer who rejects any and every authority, for fear lest it may enslave the mind.

It is true that our churches and the dogmatic tenets of the churches are full of errors, and religion as generally taught, is defaced with superstitions. But the freethinker who casts away religion is like the bear of the hermit. To drive away the fly on the face of his master, he crushes his head and kills him.

You hate oppression and yet you make your children slaves of their follies. You love liberty but you shut the door to that enlightenment without which liberty is impossible. The Churches with all their errors are by far superior to the wiseacre who destroys only, but does not build!

It is not the churches you should oppose, but the errors of the churches; it is not religion you should destroy, but the superstitions of religion! If you undermine the basis of ethics in the name of Liberty, then you are the salt that has lost its savor.

The churches have repeatedly refused to be the leaders of humanity. Whereat liberal thought was called upon to shape the future destinies of man. Ye men of a liberal mind and of progressive views, ye are now expected to be the masterbuilders, to lay the foundation. But it appears that on you the word will be fulfilled again. Many are called, but few chosen. The many have again rejected the only foundation upon which the temple of humanity can be raised.

Our people will pay dearly for the errors committed by the blind guides. The cornerstone of man's welfare is religion, and if man will live, he must take care of his soul. Tear down religion, neglect the most precious treasures that are entrusted to you, the souls of yourselves und your children, and you will reap the destruction which you deserve. The masses of our nation seem to be blind to the truth. They follow the false prophets. But let us not despair, for in the end our people will bethink themselves of the right path. Then religion shall be raised up again and the rents therein shall be closed. Then the prophetic word will come true again: The stone which the builders rejected, the same is to become the head of the corner!

THE MOTE AND THE BEAM.

THE duty of the church and of all religious congregations is to preach morals. Religion should be man's guiding star through life. Religion, therefore, must give in great and plain outlines a conception of the world in which we live, and teach us how to regulate our conduct in agreement with the facts of life, for the benefit of ourselves and our family, our nation and humanity. If the church ceases to preach morals, or if it preaches wrong morals, its influence becomes injurious to the members of its congregation and dangerous to society.

As a matter of fact we must acknowledge that the churches have done much in preaching morals; they have accomplished great things in preserving communities and making our men and women strong in enduring the tribulations of life and resisting its many allurements. Let us take one example only which brings home to us the wholesome influence of religion. Let us read a description of the Puritans as they are characterized by an impartial historian:

"The Puritan was made up of two different men, the one all penitence, gratitude, passion; the other proud, calm, inflexible. He prostrated himself in the dust before his Maker. But he set his foot on the neck of his king. In his devotional retirement, he prayed with convulsions, and groans, and tears. He was half maddened by glorious or terrible illusions. When he took his seat in the council, or girt on his sword for war, these tempestuous workings of the soul had left no perceptible trace behind them. But those had little reason to laugh who encountered him in the hall of debate or in the field of battle. These fanatics brought to civil and military affairs a coolness of judgment and an immutability of purpose which some writers have thought inconsistent with their religious zeal, but which were in fact the necessary effects of it. The intensity of their feelings on one subject made them tranquil on every other. One overpowering sentiment had subjected to itself pity and hatred, ambition and fear. Death had lost its terrors and pleasure its charms. Enthusiasm had made them stoics, had cleared their mind from every passion and prejudice, and raised them above the influence of danger and corruption."

The virtues of the Puritans, it cannot be disputed, preserved them in the calamities that had been visited upon them in their old country; they pointed out to them the way to their new home, and when they arrived in the Mayflower on the shores of the new world, it was these virtues again that made their enterprise successful. Many of the pilgrims died of cold and hunger; yet the little colony of emigrants did not despair, and finally they triumphed in spite of every adversity. The virtues which preserved them, which were the cause of their final success, what were they but religious?

Compare the history of the pilgrims to the fate of those noblemen who landed in Virginia under Captain Newport in 1607. Why was their enterprise a failure? Because they lacked the energy and endurance, the patience and self-possession of the Puritans. They had no religion to teach them these virtues, and they came over in the hope of becoming rich without work. They expected pleasures and found innumerable hardships. They sought happiness and were soon confronted with dangers and disasters which they had neither the courage nor the strength to resist or to overcome.

Why is it that among all the colonies planted on our shores the most flourishing were those founded by religious exiles?

Religion is a great power, and the religious instinct will do great work, be it for good or for evil. We know that the churches made mistakes; we know that, through persecution, they induced people to commit most heinous crimes, that they opposed, and oppose still, the progress of science. And since they suffer our conception of the world and life to become distorted, their moral preaching is in danger of leading astray. We object to their oppression and protest against the fetters with which they shackle our minds and endeavor to tie us down to certain traditional errors which they regard with reverence.

The most violent assailants of the churches are certain freethinkers who declare that all religion is superstition and that religion must be killed like a wild beast, a turbulent hyena; we must rid ourselves of religion as if it were obnoxious vermin or a lingering disease. These freethinkers, as a rule, look upon clergymen as imposters and hypocrites and are in their turn by faithful believers regarded in a similar and not a more favorable light. Most of these freethinkers are as honest as their adversaries, yet, like them, they are one-sided. They step forth and say to the people:

"Why do you allow yourself to be imposed upon by religion? Religion is an invention of kings and priests to keep the masses of the people in subjection. Religion is a humbug and the rules prescribed by religion need not be followed. Live as you please and take out of life whatever pleasures you can get. That is the sum and extract of all philosophy."

The narrow orthodoxy of the churches is the mote in the eye of our clergy. How many of our ministers feel in duty bound to impress the dogmas of their sect upon their congregation and forget the main duty upon which all their work should abut, viz., to preach morals, to make of the souls that are entrusted to their care, characters strong enough to face the adversities of life, to endure troubles, and to resist the dangers of temptation. Clergymen generally forget that the most important moral rule is the love of truth, and truth must be judged by scientific evidence, not by its agreement with, or disagreement from, the tenets of their creed.

Such is the mote in the eye of the church. But the beam in the eye of destructive freethinkers is their unqualified contempt of religion. They have become blind to the importance of morality, and the preaching of morality. Not as if they were immoral themselves, or intended to spread immorality among our people, which as they well know would lead us into speedy ruin; but because the beam in their eye,—their contempt of all religion,—has made them blind to the fact that their own morality is a treasure inherited from their religious forefathers, a treasure that will soon be wasted in the coming generations of their irreligious descendants.

Churches have faults, and some of their faults are

most grievous. Their dogmas are untenable unless a free interpretation be used. Yet their ethics, although wrong in some points, is upon the whole right. It is the ethics of the churches that kept them alive. It is the virtues of religious citizens that make colonies and nations thrive. Iconoclasts are right when protesting against the faults of the churches, against the false pretensions of religious authorities. But they are wrong when they attempt to destroy the institutions created for, and devoted to, the purpose of preaching morals.

The creed of the pilgrims was wrong in many respects; yet it was right in so far as it made of simple-minded men heroes, who could become the fathers of a great nation of liberty. The fathers were in their way freethinkers also; but they were constructive freethinkers, not destructive. They found some flaws in the religion that was taught them; yet they did not therefore throw away the whole ideal of religious life. They effaced the flaw as well as they understood to do, and preserved their ideals.

Life is a school. All of us are given a work to do. Among the scholars in the school of life, there are two: the orthodox believer and the agnostic nonbeliever. The one is plodding quietly along and tries to solve the problem given him; yet he makes mistakes. The other does not try to solve the problem, he thinks that the problem is insolvable, and seeing some blunders in the lesson of his schoolmate, attempts to erase the latter's work entirely. It is well that the agnostic should call attention to the errors of the orthodox, but the attempt to cast away that which is true and good in religion together with its errors cannot be recommended. There is a mote in the eye of the one, and the other, presuming to be the corrector and leader of his comrade, is not aware of the beam in his own eye.

Liberalism will never succeed in conquering the orthodoxy of the churches unless it offers something better than the ethics of ecclesiasticism. Liberalism must teach us morals, and its morals must be better than those of the church, its sermons must be based upon scientific truth, and must apply to the practical issues of life. Liberalism should be positive and constructive, not negative and destructive. It is true that it was necessary to destroy the old errors, but now we have done with tearing down and we intend to use the empty space to build upon it greater and nobler ideals.

Let liberalism be more than hostility toward antiquated traditions; let it cease to preach hatred of religion; and liberalism will rise in its grandeur to be the religion of mankind.

SUPERSTITION IN RELIGION AND SCIENCE.

It is not an uncommon attitude among freethinking people to see all the glories of science in its ideal perfection, and to discredit religion with the worst deficiencies it ever possessed and thus to identify it in this contrast as superstition pure and simple. This attitude is wrong, yet it is the natural consequence of religious dogmatism. A dogmatic believer when comparing science and religion, is apt to recognise the evolutionary element in science and to ignore it in religion. He knows very well that the present state of science is not its aim and end, our present knowledge is not absolute truth and the full realisation of the scientific ideal. Yet he is inclined to consider his religion as absolute and as a model of perfection. It is not natural that the unbeliever who sees the faultiness of the present religious conceptions, condemns religion itself for the sake of the errors of religious people.

But is not the dogmatic view of religion a plain and obvious mistake. Have not the dogmas, in spite of all the attempts to make them rigid and immutable, changed constantly and are they not even now almost visibly changing in all the churches? Religion is as little absolute truth as is science. Both evolve and they must evolve, both grow and develop and they develop together. A progress of science is always a prophesy for a progress of religion. And this evolutionary power, far from being an evil, is their life. Without the faculty of growth science as well as religion would be dead.

During the last few centuries all the sciences have been revolutionised by new discoveries, just as our civilisation has been modified by the many inventions made in all branches of life and labor. It is but natural that religion also should be revolutionised and based upon other principles than heretofore. This will be accomplished whether we champion or oppose the new view of religion, for it is the outcome of an evolutionary process in the growth and development of mankind.

The fact is well-established and yet little appreciated that science has just as well its orthodoxy as religion. Science in former centuries was just as dualistic as religion. And the history of civilisation is the slow process by which man frees himself from superstition. Superstition is not necessarily a religious error. By far the most numerous superstitions are scientific superstitions. Superstition is the assumption of an error as if it were an axiomatic truth; and one of the most important causes of superstition is the dualism of former centuries. Those who cherished their superstition as absolute truth assumed the name orthodox, viz. the men whose view is correct. They denounced the heterodox as revolutionists who destroyed science as well as religion.

Copernicus, Kepler, Galileo and other great scientists were to the scientists of their era heterodox, just as Luther was denounced as a heretic and infidel by the church. Socrates was executed because he was said to be irreligious, and Christ was crucified for blasphemy.

If to-day a scientist would try to establish a new—although correct—explanation of certain natural phenomena, which appeared to be contrary to the present views of his colleagues, it is certain that his theory would for a long time be rejected and ridiculed. La Marck and Darwin have experienced the truth of this fact. Only by great efforts did they and their followers overcome the old superstition of the orthodox pharisees of science.

The superstition of former ages, the erroneous dualism which boasted so much of its infallible orthodoxy, was not only an attribute of the religion of the middle ages but also of its philosophy and science. It is but a few decades since physiology got rid of the dualistic view of a life-principle, or vital power. Even to-day our chemists speak of organic and inorganic chemistry, as if two different kinds of elements existed, the living and the dead. This view and its whole terminology are but scientific superstitions.

It is not the place here to point out why the path to truth necessarily leads through errors. Nor can we here explain at length how the errors of old—far from being absolute errors—were the germs of truth. They contained golden grains of truth, and the faithful enquirer winnowed them until the grain was separated from the chaff. Thus Copernicus and Kepler were guided in their great discoveries by the old superstitious notions of the Pythagorean philosophy. They believed *a priori* in the harmony of the spheres.

Also another fact can only be hinted at: Humanity does not consist of single individuals but forms one great unity. The single individual is merely the representative of the ideas of his age, which are the results of a long process of evolution. This will easily explain why certain ages bear a certain uniform character.

There are, no doubt, exceptions. Some men are greatly in advance of their times and some lag behind. But such exceptions confute our argument as little as cases of atavism overthrow the theory of evolution.

I argued with many different persons upon the topics of religion and science, and found that apart from a difference of definitions, fundamentally they held almost the same opinions. The atheist and the monotheist have different definitions of God. The former rejects, the latter accepts, the idea of God, but *de facto* both agree much more than they are themselves aware of. The Roman Catholic priest of to-day and Robert Ingersoll are more alike in their *philosophical* views than is generally supposed, but we must eliminate the differences of their terminology and translate the language of the one into that of the other. A free-thinker of to day differs much more from a free-thinker of mediæval times than from an orthodox believer of to-day; and a Lutheran clergyman differs in the same degree from Luther himself. What Lutheran clergyman would throw his inkstand at the devil or order a misformed babe to be drowned, because it may perhaps be a changeling? What Calvinist of to-day would burn a man who had a peculiar idea of the Trinity of God. The shortcomings of religious men are not errors of religion; just as the *ignis vitæ* was not an error of science. Errors and superstitions are errors of men and of their times, and our own time has likewise its full share of them. The scientific and the religious spirit is constantly endeavoring to free humanity from its many errors.

Taking this ground, I fail to see why religion should be identified with the errors of the past and science credited with all the great ideals of the future. Why shall not religion just as well as science be freed from the shackles of superstition? Absolute truth never existed either in religion or in science. Scientific definitions and religious dogmas have changed from century to century, but the religious spirit and scientific spirit remained the same. The scientific spirit is characterised by a pure love of truth, and true religiosity means man's consciousness of being in unity with the whole Cosmos—whether it is called the All or God, Brahma or Nirvana or even Nought. The religious sentiment is a powerful factor in every human being. It prompts us to live in accordance with what we call ethics, and by it our ethical instincts must be explained. The professedly irreligious possess this religiosity sometimes stronger than those who profess a certain religion. Call it other than religion, if you please, but the rose would be a rose with any other name. In this sense Schiller said:

> "Which religion I have? There is none of all you may mention
> That I embrace; and the cause? Truly, religion it is!"

The religious spirit and the scientific spirit are so much in harmony that one cannot exist without the other. All the prominent men of science were sincerely religious—they were not orthodox; how could they be so narrow-minded if they were to be the representatives of progress? They were intoxicated, as it were, with their zeal for truth. They felt that the heart-blood of human progress was throbbing in their veins. A greater power than themselves had taken possession of them. They were conscious of working and suffering for a great cause, in comparison to which their individual loss and anxieties were but fleeting trifles. The same can be said of great artists. Such sentiment is the true religious spirit of which Goethe speaks :

> *Wer Wissenschaft und Kunst besitzt,*
> *Der hat auch Religion;*
> *Wer aber beide nicht besitzt,*
> *Der habe Religion.*
>
> The man who science has and art,
> He also has religion.
> But he who is devoid of both,
> He surely needs religion.

And this leads us to another point. Science is the privilege of the few, but religion may be had by the masses. Not everybody can be a scientist, but everybody can be and should be imbued with the true religious sentiment. Religion is not a deep philosophy, it does not take the profound learning of a scholar to recognise that the individual is but a part of a greater whole. Every child can know that; and every child should know it, not by being taught so at school, but by seeing its parents act accordingly.

A true scientist and a great artist conceive that all natural phenomena are but so many instances of the HAN KAI'EN. Nature is one and the same everywhere. Science and art are based upon this truth. Accordingly, every true scientific man, every great artist must *eo ipso* be possessed of the right religious spirit. However, those who cannot intellectually grasp this truth, must needs be religious or they will sink below the level of the savage and the brute.

What we want is religion for the masses; not orthodoxy to make them bow down and worship idols, but, a religion that makes the individual feel himself the representative of a higher power, of his community, of his nation, of humanity. A nation in which the masses are religious in this sense will be truly republican, for every citizen will be a representative of the sovereignty of the nation—of the sovereignty with all its prerogatives as well as its obligations.

THE QUESTIONS OF AGNOSTICISM.

THERE are questions that rise unasked; they obtrude upon the human mind and cannot be banished, because they lie in the nature of things. These questions so long as they remain unanswered, will cause an unrest in our soul, a spiritual thirst that can only be quenched by the spiritual waters of life—by truth and by a joyous submission to truth; they will appear as a strong and unsatisfied yearning for something that will afford help in time of need, and that shall bring light when we sit in darkness.

This dearth of peace of soul has created religion, it has created the great cosmic ideal of mankind, the idea of God as the Lord who made heaven and earth, who will be our keeper and who will preserve our soul. This dearth found expression in David's psalm:

"As the hart panteth after the water brooks, so panteth my soul for thee, O God.

"My soul thirsteth for God, for the living God; when shall I come and appear before God?

"My tears have been my meat day and night, while they continually say unto me, 'Where is thy God?'"

Our world-conception has greatly changed since David's time, and together with it our religious views have been modified. But the same yearning obtains for peace and soul; because according to the nature of things the same questions rise again and again, sternly demanding to be answered.

The same anxiety as in David's psalm pervades a communication presented to me some time ago, which in accordance with the spirit of our age formulates the thirst of the soul for a satisfactory solution of the eternal problem of life in definite queries. The letter is characteristically signed "Agnostic," and reads as follows:

"Will you kindly answer the following questions? The future of religion depends, it seems to me, on the answers given.

1) Has the universe an ethical purpose or tendency?

2) Have we any reason to believe that anything corresponding to human life, feeling, or intelligence, exists now in other parts of the universe, or will come into existence again, after the destruction of the earth?

3) Are there any grounds for hope that pain will be diminished and pleasure increased, to any great extent, in the future of humanity?

4) According to the doctrine of Evolution, will not the earth and the whole solar system, in the distant future, become, once more, a mass of homogeneous vapor, destitute of life, as the term 'life' is generally understood?

5) If the universe is an infinite machine, which mercilessly crushes between its cogs, not only the individual, but eventually the race, must not the contemplation of the universe awaken feelings of melancholy and despair in the human heart? And are not such feelings destructive to religion and ethics?"

* * *

This is an age of eager research. Wheresoever we look, we find unanswered questions; and many people shrug their shoulders in despair, because they do not expect that these questions will ever be answered. Such people call themselves agnostics.

There are three attitudes of agnosticism. There is, first, the agnosticism of indifference. This is the position of those who do not wish to be bothered with questions which they feel incompetent to answer and which they generally care nothing about. The agnosticism of indifference is passive ; it is a philosophy of indolence, which boasts of depth where because of its own littleness it has not found bottom.

The second kind of agnosticism is an agnosticism of despair. It is the agnosticism of "world-pain," and has been characterised by Heinrich Heine in the following lines :

> " By the sea, by the desolate nocturnal sea,
> Stands a youthful man,
> His breast full of sadness, his head full of doubt.
> And with bitter lips he questions the waves:
>
> ' Oh solve me the riddle of life!
> The cruel, world-old riddle,
> Concerning which, already many a head hath been racked.
> Heads in hieroglyphic-hats,
> Heads in turbans and in black caps,
> Periwigged heads, and a thousand other
> Poor, sweating human heads.
> Tell me, what signifies man ?
> Whence does he come ? whither does he go ?
> Who dwells yonder above the golden stars ?'
>
> The waves murmur their eternal murmur,
> The winds blow, the clouds flow past.
> Cold and indifferent twinkle the stars,
> And—a fool awaits an answer.*

There are men of great talents who have grappled with the questions of the day, yet have failed to solve them. They feel their labors lost and their energy, spent in thought, wasted. But because a genius has failed to solve a problem, is it really absolutely insolvable? And if it is absolutely insolvable, would it not in that case be a pseudo-problem ? A pseudo-problem is a question which is formulated on a misconception of facts ; it is unanswerable because it is misstated. The problem of existence is unanswerable perhaps, not because the world is out of joint, but because the position of the questioner is wrongly taken.

The third kind of agnosticism is the agnosticism

* Translated by Emma Lazarus.

of science. We might call it with equal appropriateness either the agnosticism of ignorance or the agnosticism of wisdom. For it is a wise confession of ignorance. This confession is not made in general terms, that science is vanity and that all philosophy is trivial. Such general statements have no meaning, except that they place the sage and the fool upon the same level. The agnosticism of ignorance is the agnosticism of science. It is an active attitude of agnosticism. It states definitely a special ignorance of ours, and formulates it in exact terms.

The statement of such a specified ignorance is called a problem, and although it may sometimes be extremely difficult to solve a problem, the agnosticism of science never despairs of a final solution. On the contrary, every problem is formed with the outspoken hope that in the end, it will be solved. The history of science is a continuous conquest of the hydra-like growing heads of the agnosticism of science.

* * *

There are certain questions—viz., the moral questions—the nature of which is such as to demand an immediate answer. "What are the rules of conduct? and what are the notions according to which we have to form these rules of conduct?"—are questions that are urgent. We live and act; and we cannot wait until science has settled all the problems the solution of which in this or in that way might influence our actions. We have to act as best we can. The notions in agreement with which our whole demeanor has to be regulated, are called "religious"; and it is natural that religious ideas through their extraordinary practical importance are of an extremely conservative nature. They are laid down as the most sacred possession of mankind, the holiest heirloom received from our ancestors. This conservatism is natural, but it will become dangerous if it prevents the revision of religious ideas through the best, and truest, and most earnest critique that can be furnished by science. It will become detrimental if it produces thoughtlessness, and makes a generation accept without critique whatever it has been taught to believe.

It lies in the very nature of religious problems that they must be solved again and again. Every one of us has to solve them for himself as best he can. It may be stated parenthetically that most religions are creeds; but they need not be creeds and the Religion which we advocate is the Religion of Science.

The questions proposed by Agnostic are in their nature religious questions, and we answer them very briefly as follows :

1) "Has the Universe an ethical purpose or tendency?"

If this question is to be answered by Yes or No, we should say, Yes—the universe has an ethical tendency. But it must be borne in mind that this way of putting the question is incorrect. We should ask whether the universe has any definite tendency, or whether it has no definite tendency whatever, without calling its tendency either moral or immoral. If the universe had no definite tendency it would be no universe, no unitary world, no cosmos, but a jumble of incoherent events, a chaos, a labyrinth of heterogeneous things, a confusion without rhyme or reason, without law or order. Our answer to this first question is, that the universe *has* a definite tendency, and morality means agreement with this tendency.

2) We have reasons to believe that on other plan-

ets and in other solar systems, there is something corresponding to human life, to feeling, and to intelligence. For philosophical considerations teach us, and science corroborates it, that the evolution of the human race, the feeling of animal life, and the intelligence of rational beings have developed with necessity upon earth in rigid accordance with natural laws. Is there any doubt that the same conditions in other parts of the universe will produce the same results, and similar conditions similar results? When we analyze the stars with the assistance of the spectroscope we find there the same material elements as upon the earth. Can there be any question as to our finding everywhere the same laws and the same tendency of evolution? Other races on other planets may have very different constitutions; winged animals of the air or swimming animals of the sea, bipeds or quadrupeds, mammals or insects, carnivorous or herbivorous, or any other kind of creatures might develop into thinking beings; yet it is certain that among all rational creatures, there would be at least in all fundamental features the same logic, the same arithmetic, the same mathematics, and above all the same logic of action, viz., the same ethics.

3) There are grounds for hope that pain will be diminished in life and that the nobler and more refined pleasures will be constantly increased. But considering that pain is either the result of unsatisfied wants or due to some other disturbance in life, we must bear in mind that the creation of new wants which arises through progress, will produce new pains to the same degree as it will produce more refined and nobler pleasures.

Are we not sometimes too weak-hearted with regard to our pains? Are not the causes of our woes mostly of a trivial nature? Look at them from a higher standpoint and they appear like the baby's tears over a broken doll. And if they are not trivial, if they are not the woes of the individual, but of the aspiring race, are they not far from being merely lamentable? Are they not in such a case sublime? Are they not transfigured by their sacred purpose, and must they not appear as grand as are the struggles, the anxieties, and the sufferings of a hero in a tragedy?

Let us consider pleasure and pain not from the standpoint of sentimentality but from the higher standpoint of ethics, where the individual as such disappears, where the individual's worth is measured according to his breadth of mind, and where life is valued not according to the pleasures it affords, but according as it contains more or less of those treasures that "neither moth nor rust doth corrupt."

As to the fourth and fifth question, we should say:

This planet of ours together with our solar system may, and we have indeed reasons to believe that it will, break to pieces. Yet the conditions which produced not only our solar system, but also mankind and human civilization, will not cease to exist. They will continue to exist and will produce, in fact they are constantly producing, new worlds out of the wrecks of the old broken ones. If a man dies, we lament the loss; we weep for the friend, the brother or the father. But the loss is not so much his; it is ours. If our world breaks to pieces it will be a loss—a lamentable loss. But will it be a loss to mankind? It will be a loss in the universe, which, however, as we can fairly suppose, will be made up by other gains.

The universe is *not* "an infinite machine, which

mercilessly crushes between its cogs not only the individual but eventually the race." The universe is infinite and inexhaustible life. Whatever life of organized beings, of individuals, of entire races and of entire solar systems may disappear in one part, there is a probability, practically amounting to certainty, that in other parts new life will originate to compensate for it.

Life on its highest stage means action and action means performance of duty. Man is an ethical animal, which means that he has come to understand certain important features of the tendency prevailing in the universe. It is the performance of duty in past generations which has raised mankind to its present eminence.

The world is throughout a field of ethical aspirations. If our life ceases, if our planet breaks to pieces, the immutable laws of nature will remain the same. Humanity may be wiped out of existence, but those realities which created humanity and in consonance with which man's ethical ideals have been shaped will remain. We read in the New Testament that Heaven and earth may pass away, but the word of God abideth forever. The Religion of Science recognizes the truth of this biblical verse, although it does not accept it in the narrow interpretation of theistic theology.

THE BIBLE AND FREE THOUGHT.

At present there are two distinct views concerning the Bible, viz., that of the so-called orthodox, and that of the irreligious radical. Those advocating the former view believe that the Bible was revealed by divine inspiration and communicated word for word. They declare that it contains nothing but truth,—absolute truth. The advocates of the latter view consider it a book full of paradoxes and contradictions. They ridicule it as the *non plus ultra* of superstition and the very basis of bigotry.

Both parties are in error. The Bible although not dictated by the Holy Ghost *verbatim*, is from a human and secular standpoint the grandest and sublimest book we have. Compare it with the sacred books of other nations, with those books which are the old store-houses of ethical, religious and mythological ideas. Compare it with the *Koran*, with Hesiod's *Cosmogony*, or the Völuspa of the Northern Edda, or the *Zend-Avesta*, or even the *Vedas* and the *Buddha Gospels*. What impartial judge would not give preference to the Bible?

Goethe found in the Bible an invaluable store and an inexhaustible mine of poetry; he ranked it far above Homer. Read the passage in Humboldt's *Kos-*

mos, where he expresses his admiration for the Hebrew literature and more especially the poetry of the psalms!

The sacred books of all nations, and particularly the Bible form the basis of our modern ethics. That the Bible should bear traces of the times in which it was written, is quite natural. But it also points far beyond its time, in that it contains germs which have developed into a higher ethical culture. It is this that gives to the Bible its value.

The Bible, when regarded from the standpoint of narrow bigotry, becomes a tissue of almost unexplainable absurdities. How many things, which can be explained by the ideas and manners prevalent in those times, must now appear incongruous. No matter how much the irreligious and flippant scoffer may differ from the bigot in his ultimate opinion concerning the Bible, his view nevertheless coincides with the latter's in that they both guage the Bible according to the same standard. Both demand proofs of absolute truth; and because the infidel does not find them he deprecates it and ridicules the pretensions of believers. Both the bigot and the scoffer lack scientific insight.

If we consider the Bible from the standpoint of the severest and most radical criticism, we shall only learn to prize it all the more, on account of its poetical treasures and on account of the valuable evidence it affords of the growth of religious, ethical and philosophical ideas.

From this standpoint of careful and earnest scientific investigation the Bible will be read with the greatest pleasure and edification.

We prize our old legends of fairies and witches, heroes and ogres, of the shepherd boy who slays the giant and becomes a king, but we are blind to the beauty of the story of David and Goliath. And why are we unable to appreciate its charm? Is it not because, when we first read it with our teacher, the human features of the story were ignored? They were purposely thrown aside and something superhuman, something awe-inspiring was wrongly substituted; and this made the whole tale unintelligible to the child.

The Bible if not distorted by narrow-minded bigotry is a rich mine for every one. The child's love for stories is satisfied, the historian finds records which are of the greatest importance for our knowledge of the patriarchal era of mankind, its customs and habits, its beliefs and superstitions, its laws and its culture. And above all, those who want to found their actions upon a firm basis of rules and principles, who aspire toward religious or ethical ideals, will find the most fertile fields in the books of the Bible, if they search in the right spirit, prejuduced neither by credulous acceptance nor flippant rejection of all their contents.

The Old Testament is one of the strongest supports of free thought, and the words of Christ are so full of truth and righteousness that they have rung through almost nineteen centuries and have not as yet lost their power. They have been wrongly interpreted, they have been scoffed at and ridiculed, they have been criticised and condemned, but they survived nevertheless, and will live on in the ethical development of humanity. The radical freethought of the Bible is perhaps not understood by those who say "Lord, Lord," who read and worship the letter and lose sight of the spirit.

Mr. Salter, the well known lecturer of the Society for Ethical Culture in Chicago, speaks in *The Chris-*

tian Register (Jan. 19, 1888) of the significance of Jesus for our time. He says: "The charm about the name of Jesus is that he dared believe in something different from what he saw about him. He loved justice in his soul, but with his eyes he saw injustice."

Christ's word, "Ye resist not evil," is a lesson to the human race which people even to-day have not yet understood. We are still prone to obey the old rule: "An eye for an eye and a tooth for a tooth." If one does injustice to another, this other thinks the best remedy is for him in his turn to do another injustice. It is almost an unwritten law of our social code "to render evil for evil and railing for railing." If the monopolist oppresses the workingmen, the trades-unions expect to help themselves by committing a similar injustice. We must be educated to "a perception," as Wheelbarrow says, "strong enough to see that freedom to oppress others is not freedom." It will perhaps take some centuries for society to learn that the wrong-doer injures others and himself still more. He who seeks revenge by retaliating does not right the wrong but aggravates it. He intends to restore justice and increases injustice.

There are but few who can distinguish between an honest fight with their adversary and a hateful persecution of their enemy. The former is our duty, the latter is deplorable, and if done in a cowardly manner with the help of lies and slander, it is even despicable. So long as we stick to the old rule of rendering evil for evil, every evil will beget a new evil. But if we let it alone, if we fight our struggles honestly without bearing any hatred toward our adversary, evil will be exterminated.

The real Christian is not he who believes the marvellous stories told in the Bible, but he who acts in accordance with the teachings of Christ, which finally must be recognised as true in their spirit and humane in their nature. They are right and correct and will outlast the worldly wisdom of retaliation. They will come to be recognised more and more, not only as noble and sublime from the ideal point of view, but also from the lower standpoint of practical prudence.

We would therefore call the attention of the freethinker and of the bigot to the Bible. The one will find in store for him treasures of most radical thought, love of justice and truth, which he did not expect, and the other will learn that Christ was different from what he is generally represented in the orthodox pulpits. Our modern ethical civilisation is evolved from the biblical teachings and we have not as yet been able fully to comprehend all the ideas embodied in them, nor to realise them in actual life. Mr. Salter in the above quoted article says: "Religion must inspire to personal and social reform. That is the only thing that is religion in the modern world. All else is the tradition of an earlier time, when justice and judgment were committed to other hands than man's.".... "We cannot pray for justice any longer. We have to do it. We cannot say, Thy kingdom come. We have to obey the God who commands us to create it."

If any one who claims to be a teacher of free thought and ethical progress, disdains the prophets in the Old Testament, or the Doctrines of Christ in the New Testament, if he scoffs at his followers, the Apostles, Paul, Augustin or Luther, because they were in many respects not so far advanced as we are now, he seems to me like an engineer who foolishly prattles about the stupidity of Watt and Stephenson or other great inventors because their engines were poor in comparison to the engines of to-day. An engineer of such stamp will not become an inventor. Due reverence for and appreciation of the merits of the past is the only foundation on which a truly grand future can be built.

Radicalism is needed in our churches and our clergy should know that free thought—in its best sense—can never destroy religion, but on the other hand religion is wanted among our freethinkers. They should know that true religion is the most radical power of a consistent free thought which in constant opposition to narrow-minded bigotry leads humanity onwards in the path of progress.

FAITH AND DOUBT.

The value of scepticism was the subject of a discussion in a club consisting mainly of scientists, lawyers, and business men. And it was a strange fact that almost all the speakers glorified scepticism as if it had been the cause of all progress, as if the human mind reached the climax of perfection in Doubt.

This attitude, it appears, is based upon an erroneous conception of the function of doubt, and it is now so prevalent partly because the terms doubt and scepticism are often identified with any denial of certain religious beliefs, and partly because agnosticism, which despairs of a definite solution of the fundamental problems of philosophy, is at present the most prevalent and fashionable world-conception.

In the addresses made, it was maintained that all success in life was due to doubt. An able business man had doubted the propriety of the prevalent methods of distribution in the meat-market; and Charles Darwin had doubted the truth of the biblical account of creation, and lo! what were the results? The former created an establishment which made meat cheaper all over the world, and the latter wrote "The Origin of the Species" and "The Descent of Man." One of the speakers defined doubt as the faith of a man in himself and in his ideals, contrasting it with a blind faith in dogmas. But it strikes us that this view of doubt and scepticism is, to say the least, misleading. Doubt, real doubt, is unable to produce any results. The man who has a faith acts according to the faith that is in him. But the man who doubts is like Buridan's donkey who hungers between two bundles of hay so long as he remains in the agnostic state of not knowing which bundle should be eaten first.

It was maintained, likewise, that the times of scepticism had been the times of progress. This is true only if scepticism be identified with active thought. Goethe said, that the epochs of strong faith alone had been the periods of a strong activity, of progress, of creative thought, fertile with ideas and deeds. It is not true that Mr. Armour's doubt produced the new methods of the distribution of meat, it was his faith in the new methods and not his doubts as to the old methods that produced progress. The negative element of doubt, important though it may be as a transient phase in the growth of our ideas, is not so important as the positive element of a new faith for the creation of great things. It is most probable that the new faith in the truth of the evolution theory developed in Darwin's mind long before his old faith had broken down, and it is not impossible that for a long time he did not even realise the full extent of the conflict between the old and the new faith. Success after all is always due to faith; and doubt is nothing but a state of suspense in which a new faith is struggling with the old faith, and only lasts so long as both faiths are sufficiently equal in strength to paralyse each other.

The aim of doubt is always its annihilation. Problems tend to be solved and the end of doubt should be their settlement. But we were told by an encomiast of scepticism, that "theories and dogmas vanish in a clear and keen cut mind before doubt, even as mists before the morning sun." However if the old theories are not replaced by new and better theories,—better because they are truer,—it would seem as if we should rather compare the state of doubt to the mist. For if we are surrounded with a dense fog we cannot see, and only so long as we are in doubt do we answer "Alas! I know not."

It is strange that the doubt of this same eulogist of scepticism is not at all a state of not knowing. When he attempted to explain the actual advantages of doubt he became inconsistent with himself. As soon as he tried to describe his doubter's "hope eternal" it was noticeable that doubt became simply a wrong name for the opposite of doubt. What he calls doubt is actually a new faith. His "doubter mourns not, not as one without hope," for he positively knows that "we live and die by laws as inevitable, all working toward a unity of completion" and "Nature makes no blanks," and death has also its place in nature. "It is death that weaves a crown for birth and life."

A new faith is dawning on the intellectual horizon of mankind; and whether the new faith should be considered as preferable to the old faith has, to the large masses of our people, not as yet been decided. Hence the prevalence of doubt. This prevalent state of doubt is unquestionably the harbinger of better days, it is a sign of progress, it promises life, and growth, and evolution. But let us not make doubt the aim and end of thought. Our ideal is not the despair of an eternal scepticism, but the great hope of a new, of a better and a truer faith.

THE HEROES OF FREE THOUGHT.

Who are the heroes of free thought? Those who smile at religious sentiment and think that "religion is good for the masses while the educated naturally stand above any religious emotion"—or those who struggle and yearn for truth, who suffer for it and advance slowly, but earnestly, on the path of human progress? The former may be more advanced in refinement, knowledge and worldly wisdom, but the latter only are the heroes of free thought. Such men were Giordano Bruno, Spinoza, Luther, Lessing, Hume, Kant and others, and it is noteworthy that almost all of them were not only from childhood earnestly pious, but that they also came from families where religion was more than the mere observance of ceremonial rites.

Let us confine ourselves to the best known of such characters. David Hume was a Scotchman, whose ancestors were, as are all the old Scotch people, very devoted Puritans. Kant, also, was of such Puritan Scotch origin, and we know that his mother was a devout Christian.

Spinoza was a Jew. His parents left their home in Spain for Holland, in order to remain faithful to the religion of their ancestors. They might have comfortably remained in Spain if they had abjured their belief and turned Christians. The religious spirit of Spinoza's writings is fully appreciated even by his adversaries, and he showed this religious spirit in practical life when, for the sake of truth, he scorned the terrible curse of the synagogue, in the teachings of which he had been educated.

Luther's faith and love of truth is an historic fact. He was a hero of free thought, which his contemporary, the great Pope Leo X., was not. Pope Leo was a free-thinker of the modern stamp. Luther was a firm believer, Leo was an unbeliever. Luther had faith in God like a child. Pope Leo was unhampered by any credo and at the same time was a protector of art and a promoter of humanitarianism. He did much for the Renaissance in resuscitating Greek letters and Greek culture. He built the glorious Cathedral of Saint Peter's at Rome and to show his Helenic spirit he placed upon the cross formed by the four great aisles of the largest church on earth a cupola resembling the pagan Pantheon. In his heart Greek paganism triumphed over Christianity.

Compare this great Mæcenas, the free-thinker, the humanitarian, the erudite man, with the poor, almost illiterate Augustine monk. Would you then have recognised the power of free thought in the latter and the lack of it in the former? What gave to the simple-hearted believer the strength to lead humanity one great step onward, so as to gain for every man the freedom of his conscience—the Christian's liberty, as Luther called it? It was not that he believed less of the dogmatic Christianity, but that his religious faith was stronger. Pope Leo was indifferent to religion; he was a free-thinker, and, upon receiving the Peter's pence, spoke of "the profitable fable of Christ." He appreciated and understood Luther's opposition so little that he thought his preaching against Tetzel's sale of indulgences was mere jealousy of the Augustine monk's against the Dominican Order, to whom the sale was entrusted. Leo could not imagine that any one would endanger his life for the sake of conviction.

Luther very probably would have been shocked had he foreseen that humanity would advance on the path of religious free thought. He did not see so far. But it was better for him and better for the cause which he boldly defended. We, however, should learn from the juxtaposition of those two men, Leo X. in all his papal splendor and the poor monk Martin in his simple faith, that the heroism of free thought is no mere indifferent negation of religious dogmatism, but strong faith—religious faith—and confidence in truth. Let us boldly and consistently think the truth, let us speak the truth modestly but firmly, that is the spirit by which the heroes of free thought became a power and rose above their time so as to lead humanity to higher and nobler aims.

THE HUNGER AFTER RIGHTEOUSNESS.

There is a most dangerous superstition prevailing among great masses of people that morality is a good thing as an ideal, but a bad thing for the purposes of practical life. A business man who wants to succeed, it is imagined, can succeed by immoral means only. This is a superstition, for it is not true; and it is a dangerous superstition, for it leads those who believe in it and act accordingly, into ruin. Morality, if it be true morality, will lead to life, it will preserve, it will produce prosperity, and afford a noble satisfaction never mingled with regret.

The deep-rooted error that immorality alone can insure success, seems to have originated through a strange combination of misconceptions, favored by special conditions and strengthened by exceptional instances of successful impostors. Our very language betrays us into grievous blunders. We speak of a "smart" business man and understand by "smart" now the prudent, industrious, judicious merchant, and now the sagacious, deceitful trickster. Prudence is indispensable to insure success, but trickery is not. Trickery will go but a little way and, like the crooked boomerang, it will unexpectedly fly back upon its originator.

Closely connected with this vagueness of speech is the vagueness of our views of morality. Morality is too often tacitly identified with so-called goodnaturedness and with inability. It is proverbial to speak of incompetent men who are free from other gross faults as "good people, but bad musicians"; meaning thereby that they are morally blameless, yet still disqualified for the business or profession in which they are engaged. Such men are popularly called 'good,' i. e., morally good; but they are not good. They lack that moral nerve that enables us to adapt ourselves to our work; they lack that moral energy of self-discipline by which alone we can train and educate ourselves to become competent in our profession.

The negative morality of doing no harm to anybody is not as yet morality; it is, at best, sentimentality. True morality has positive ideals, and foremost among our moral ideals must be the aspiration of every individual to become a useful member of society, by contributing something to its weal and welfare. To do some work which gives us pleasure, dilettanteism in art or science, in business or agriculture, etc., is not as yet sufficient; our work must be a service to society, it must stand in demand, otherwise we cannot and ought not expect any return for it.

A certain indifference with regard to honesty easily arises from an over-prosperous condition of society. If men earn money without earnest effort; if they live in plenty, and find the resources of all departments of industry practically unlimited, they become indulgent towards the depredator who takes more than his due, and smile at the thief who nimbly skips away with his spoil. He who plunders the public treasury is not taken to account, because the loss is not so seriously felt. A country in an unusual state of prosperity is not so much in need of honesty as a poor nation, and accordingly the moral instinct, the moral sense of that country remains comparatively undeveloped. If man did not stand in need of intelligence, if he could live without thought, he certainly would never have developed brains, and humanity would still lead an unrational existence. The same is true of morality: it is developed among mankind because and to the extent in which man wants it. And we do want it indeed; we are most intensely in need of it, for so-society could not exist without it.

A prosperous nation, I say, is not so much in need of morality as a poor nation, where the struggle for existence is hard and competition is fierce. Yet the people that are not at present in such great need of morality will soon come to that need. History teaches that the moral, the industrious, the patient poor people will in time most successfully compete with the rich and the opulent. As soon as opulency has reached that degree in which the need of morality is no longer felt, the decline of a nation sets in. A crisis in her social life is impending. The downtrodden will complain of their oppressors; they will cry out for justice; and if that justice be not freely given, the whole nation will suffer for it, and the country once so prosperous will lie deserted and in ruins. Let the monuments of the great nations that prospered before us and passed away be a *mene tekel* for us to-day.

When the nation of Israel was in a social condition similar to that which, to a great extent, prevails among us now, the prophet Amos arose and lamented the moral depravity of his people. He said:

Thus saith the Lord: For three transgressions of Israel, and for four, will I not turn away their punishment. For they sell the righteous for silver, and the needy for a pair of shoes. And pervert the cause of the afflicted. They lay themselves down upon pledged garments near every altar; and drink wine procured by fines, in the house of their gods.

Amos foresaw that such a state of society could not remain as it was. He said:

And I will turn your feasts into mourning and all your songs into lamentation; and I will bring up sackcloth upon all loins and baldness upon every head; and I will make it as the mourning of an only son, and the end thereof as a bitter day.

The need of morality, its indispensableness for the welfare of the nation as well as of every individual, must at last be felt, and under the impression of this truth the prophet continues:

Behold the days come, sayeth the Lord God, that I will send a famine in the land, not a famine of bread, nor a thirst for water, but of hearing the words of the Lord.

Amos's prophecy is as true to-day—and we repeat it in this conviction—as it was about two and a half millenniums ago. There will come upon us disorder and misery, our feasts will be turned into mourning unless we are made aware of the want of honesty, of justice, of morality. The expression "the words of the Lord" in the prophecy does not signify belief in a supernatural revelation; and if it did, we do not quote it in that sense. "The words of the Lord," as we interpret the term in accordance with its context, mean the moral commands that will forever remain the substance of religious aspirations. There will arise, as Christ said, almost two thousand years ago, "a hunger and thirst after righteousness." Those who feel that hunger will partake of the blessing that in the nature of things is intimately connected with it, that will follow upon it, as the effect follows upon its cause.

Says Amos:

> For, lo, I will command, and I will sift the house of Israel among the nations; like as corn is sifted in a sieve, yet shall not the least grain fall upon the earth.

The prophecy of Amos is constantly being fulfilled in the process of the survival of the fittest. Among all the nations those alone will survive and fill the earth that are pervaded with the moral spirit. A society based upon justice will be stronger than a society in which an aristocracy oppresses the other classes of the people. A nation in which the rich devise laws to protect themselves against free competition and in which the poor are prevented from bettering their condition, carries a germ of weakness within itself and will in the end have to pay for its errors dearly. The strong will conquer and the weak will go to the wall—that is the natural law of evolution. But bear in mind that there is no strength unless it be supported by morality. The social law is a power—a power that destroys those who do not conform to it. Says the prophet:

> Yet destroyed I the Amorite whose height was like the height of the cedars, and he was strong as the oaks. Yet I destroyed his fruit from above and his roots from beneath.

Rocks are demolished by silently-working atmospheric influences. And the strongest nations perish as soon as they deviate from the path of righteousness and the spirit of progressive morality. A constant selection takes place in the struggle for existence, and humanity is sifted like as corn is sifted in a sieve.

Let us learn the truth and act accordingly, and we shall live. Let us not waver in the path of righteousness, but do faithfully some useful work in the service of humanity, lest we become like the chaff which the wind driveth away.

ETHICS AND THE STRUGGLE FOR LIFE.

THIS world of ours is a world of strife. Wherever we turn our eyes, there is war and competition and struggle. Battles are fought not only in human society, but in animal society also; not only in the animal kingdom, but in the plant kingdom; not only in the empire of organized life, but in the realm of inorganic life—between the ocean and the land, between water and air, among minerals, and among the different formations of mineral bodies, among planets and planetary systems, among suns and clusters of suns. Strife is identical with life, and struggle is the normal state of actual existence.

We can easily understand that a superficial observer of nature will feel inclined to look upon life as a chaotic jungle without rhyme or reason, in which the wildest hap-hazard and fortuitous chance rule supreme. A closer inspection, however, will show that there is after all order in the general turmoil and that a wonderful harmony results from the conflict of antagonistic principles. Nay, we shall learn that all order proceeds from the antagonism of factors that work in opposite directions. It is the centrifugal and centripetal forces that shape our earth and keep it in equilibrium. It is attraction and repulsion that govern the changes of chemistry. Gravitation throws all things into one centre, and radiation disperses the store of energy collected in that centre. And the same antithesis of hostile principles manifests itself in love and hate, in surfeit and hunger, in hope and fear.

There are many people who are not satisfied with this state of things. They dream of a paradise where there is no strife, no war, no conflict; where there is eternal peace, unmixed happiness, joy without pain, and life without struggle. Whenever you try to depict in your imagination such a condition of things, you will find that a world of eternal peace is an impossibility. The world in which life does not signify a constant struggle is not a heaven of perfection (as is imagined), but the cloudland of Utopia, an impossible state of fantastical contradictions. Should you succeed in realizing in imagination the dream of your ideal of peace without inconsistency, it will turn out to be the Nirvana of absolute non-existence, the silence of the grave, the eternal rest of death.

Natural science teaches that hate is inversed love and repulsion inversed attraction. Annihilate one principle and the other vanishes. Both principles are one and the same in opposite directions. Thus they come into conflict and their conflict is the process of life. Science does away with all dualism. The dualistic view appears natural to a crude and child-like mind. The Indian might say that heat is not cold and cold is not heat, yet the man who learns to express temperature by the exact measurement of a thermometer must abandon the duality of the two principles. Monism is established as soon as science commences to weigh and to measure. The divergence in the oneness of existence creates the two opposed principles, which are the factors that shape the world, and the encounter of conflicting factors is the basis from which all life arises with its pains and joys, its affliction and happiness, with its battles, defeats, and victories.

The world being a world of struggle, life teaches us the lesson that we live in order to fight; we must not blink at this truth, for we cannot shirk the combat. Ethics, accordingly, if it is true ethics, and practical ethics, must above all be an ethics of strife. It must teach us how to struggle, how to fight, how to aspire. In order to teach us the *how*, it must show us the goal that is to be striven for, and the ideal which we should pursue.

The progress of civilization changes the weapons and abolishes barbaric practices; yet it will never abolish the struggle itself. The struggle will become more humane, it will be fought without the unnecessary waste which accompanies the rude warfare of the savage, but even a golden era of peace and social order will continue to remain an unceasing strife and competition. You cannot abolish competition even in the most complete co-operative system. There will always remain the struggle for occupying this or that place, and the competition for proving to be the fittest will continue so long as the world lasts; and it is the plan of nature to let the fittest survive.

There are ethical teachers who imagine that the purpose of ethics is the suppression of all struggle, who depict a state of society where there is pure altruism without conflicting interests, a state of mutual love, a heaven of undisturbed happiness.

The ethics of pure altruism is just as wrong as the ethics of pure egotism. For it is our duty to stand up manfully in battle and to wage the war of honest aspirations. It is the duty of a manufacturer to compete with his competitors. It is the duty of the scholar, the philosopher, and the artist to rival the work of his co-laborers; and the progress of humanity is the result of this general warfare. Organized life from its lowliest beginnings developed higher and higher by a continued struggle; and it is not the victor alone to whom the evolution of ever higher and higher organisms is due, but to the vanquished also. The victor has gained new virtues in every strife, and it is the brave resistance of the vanquished that taught him these virtues.

There is an old saga of a northern hero, to whose soul, it is said, were added all the souls of the enemies he slew. The strength, the accomplishments, the abilities of the conquered became the spoils of the conqueror; and the spirits of the slain continued to live in the spirit of the victor, and made him stronger, nobler, wiser, better. This myth correctly represents the natural state of things, and we learn from it the great truth, that our efforts, even if we are the unfortunate party that is to be vanquished, will not be in vain; our lives are not spent in uselessness, if we but struggle bravely and do the best we can in the battle of life. Furthermore, we learn to respect our adversaries and to honor their courage. We are one factor only on the battlefield, and if our enemies existed not, we would not be what we are. We are one part only of the process of life and our enemies are the counterpart. Any contumely that we put upon them in foolish narrow-mindedness, debases and degrades ourselves; any dishonesty that we show in fight, falls back upon ourselves. It will injure our enemies, as was intended, but it will do greater harm to ourselves, for it will disgrace us; and our disgrace in that case will outlive the injury of our enemies.

Ethics teaches us that all struggle must be undertaken in the service of a higher and greater cause than our egoistic self. He alone will conquer who fights for something greater than his personal interests; and even if he be vanquished, he will still have the satisfaction that his ideal is not conquered with him. He will find successors to continue his work. His ideal, if it be a genuine ideal, will rise again in his successors and they will accomplish a final victory for his aspirations.

The Teutonic nations,—the Anglo-Saxons, the Franks, the Germans and their kin,—are, it appears, in many respects the most successful peoples in the world, because of their stern ethics of undaunted struggle to which they have adhered since prehistoric times. It was no disgrace for the Teutonic warrior to be slain, no dishonor to be vanquished; but it was infamy worse than death to be a coward, it was a disgrace to gain a victory by dishonest means. The enemy was relentlessly combated, may be he was hated, yet it would have been a blot on one's escutcheon to treat him with meanness. It was not uncommon among these barbarians for the victor to place a laurel wreath upon the grave of his foe, whom in life he had combated with bitterest hatred. There is an episode told in the *Nibelungensaga* which characterizes the ethical spirit of the combativeness of Teutonic heroes. Markgrave Rüdiger has to meet the grim Hagen and to do him battle. Seeing, however, that his enemy's shield is hacked to pieces, he offers him his own, whereupon they proceed to fight.

The moral teacher must not be blind to the laws of life. Ethics must not make us weak in the struggle for existence, but it must teach us the way to fight and must show us the higher purpose to be realized by our struggle.

Naturalists give us most remarkable reports about the degeneration of those organs and their functions and abilities which are not used. If man could live without reason, without education, language, without reason, mankind would soon degenerate into dumb brutes.

Do not attempt to preach a morality that would deprive man of his backbone. Man acquired his backbone because in the struggle for life he had to stand upright, thus to keep his own. If it were possible at all to lead a life without struggle, the backbone of man would soon become a rudimentary organ. But as it is not possible, those men alone will survive that are strong characters, that stand upright in the struggle and fight with manly honesty and noble courage. The men with a moral backbone alone are those to whom the future belongs.

Ethics must teach us how to struggle; it must not hinder us in the combat but help us. And ethics will help us. Ethics demands that we shall never lose sight of the whole to which we belong. It teaches us never to forget the aim which humanity attains through the efforts of our conflicting interests; it inculcates the lesson to do our duty in the battle of life, not only because this is required by our own interests, but because it is the law of life that we have to obey. By a faithful obedience to the ethics of the struggle for life, we shall promote the welfare of mankind and contribute to the enhancement of human progress.

RENDER NOT EVIL FOR EVIL.

God is often compared in the Old Testament to a shepherd who leads his people in the paths of righteousness; and those who truthfully obey his commands, who allow themselves to be guided by him, are called his sheep, his lambs, his flock. Christ adopted the same simile and often refers to it. In the Acts (viii, 32) Christ himself is compared to a sheep. To him is referred the prophecy in Isaiah (liii, 7): "He was led as a sheep to the slaughter, and like a lamb dumb before his shearers, so opened he not his mouth."

This comparison was sufficient to give the crown of glory to the sheep. Christians forgot that similes remain similes; that they do not cover the truth in all respects, but in one or two points only: and thus it happened that the weakness of the sheep, its simplicity, nay, its very stupidity, became an ideal of moral goodness and Christian virtue. This misconception of the true meaning of goodness received a further support in such passages as "Ye resist not evil," and "Blessed are the poor in spirit for theirs is the kingdom of heaven." Mental and physical weakness, so the doctrine of Christianity seemed to say, is a moral merit; and the principle of absolute non-resistance was seriously defended by many devout believers.

In recent times Christ's word "Ye resist not evil"

has come again into prominence through the teachings of Count Tolstoi, who not only adopted it as a practical rule of conduct but attempted to show through his example that it was possible to live up to it.

Christ's command, "Ye resist not evil," contains a great moral truth, and Count Tolstoi was taught it not through traditional belief in dogmatic Christianity, but through the hard facts of life. Having enjoyed a good education, he had become an unbeliever by his acquaintance with the so called sciences, and in his practical experiences he found himself confronted with many anxieties: care and worry for his beloved came upon him; he beheld the pale face of death; and in the moment of despair the unbeliever found comfort and strength in words of prayer.

Count Tolstoi was converted not by the sermons and representations of a subtle apologetic divine, but by the overwhelming logic of facts consisting in the moral relations between husband and wife, brother and brother, friend and friend, man and man. It was life that taught the lesson "Ye resist not evil" to Tolstoi, and his religion is a religion based upon experience.

The myths of the Saviour who came into the world from spheres beyond, contain pearls of imperishable worth. Having ceased to believe in the sacred legend, we may very well preserve the moral truths that like valuable kernels are hidden in the useless husks of dogmatism. The ethical teacher of the future while rejecting the historical fables of Christ's life with an uncompromising truthfulness, must extract the gold, purified from dross, out of the ores of the old religions.

Christ's word "Ye resist not evil" must not be misinterpreted as if it meant the abolition of all struggle and a passive submission to everything vile and low. A parallel passage, 1 Peter, iii, 8, reads as follows:

> "Be ye all of one mind, having compassion one of another, love as brethren, be pitiful, be courteous: not rendering evil for evil, or railing for railing; but contrariwise blessing; knowing that ye are thereunto called, that ye should inherit a blessing."

Christ's word "Ye resist not evil" demands the suppression of the natural tendency of retaliation. The brutish desire in man for vengeance whenever he suffers a wrong, should give place to brotherly love and forgiveness. This is a divine command. Yet divinity, as we understand the term, does not stand in contradiction to nature. Divinity is nature ennobled elevated, and sanctified. The ethics of love is divine, because it is firmly established upon the facts of life; and science, if it be not blind to the moral law that pervades nature, will find that it is true. Spinoza, whose ethics is not that of revelation, says (*Ethics*, III, 43 and 44):

> "Hatred is increased through hatred yet can be extinguished through love.
>
> "Hatred if completely conquered by love, changes into love; and this love will be greater than if no hatred had preceded it."

The evil of this world cannot be lessened by counteracting it through new evil. You cannot diminish it by committing more evils. The logic of this truth is becoming recognized in society now. Suppose that some one being in a rage, called you names. Would you stoop so low as to answer in the same tone? Would you childishly act like the bad boy saying: "You're another!" Certainly not, unless you lose your temper and do things that you will later regret.

The doctrine "Ye shall not render evil for evil,"

in this sense, will be more absolutely recognized the higher the standard of moral culture is. Yet this doctrine does not at all imply the abolition of all struggle and the suppression of combat and fight. We are too much accustomed to look upon struggle as the root of all evil, and in that case we shall erroneously expect that a world of moral life must be without competition, without war, without fight. The doctrine of non-resistance, in the sense of giving up all efforts to defend that which is right and just, is practically and morally untenable. Life in all its many phases is a constant struggle, and the ethics of life demands that we shall fight the good fight of faith trusting in the invincibility of the moral ideal.

The sentence "Ye resist not evil" is ambiguous and it appears preferable to express the truth of this doctrine in the words, "Render not evil for evil." Evil must be resisted, but not by other evils; selfishness must be overcome but not by other and greater selfishness. Therefore, by the side of the doctrine "Resist not evil with evil," let there appear the command: Do your best in the struggle for life and conquer evil, not because your own personal interests are at stake, but because higher principles are involved than the private affairs of your petty self. We must never lose sight of the truth that our struggle for existence, even in commercial competition, is fought for the progress of humanity and for an ever higher and better realization of human ideals.

Christ—that is, a moral teacher as described in the four gospels—could not possibly have meant by his word "Ye resist not evil," that doctrine of passive indolence that made of the sheep the ideal of moral perfection. For Christ himself fought and struggled, he discussed and wrangled with the Scribes and Pharisees. When he stood before Caiaphas, according to the account of John, he was smitten in his face, and although he was ready to endure another blow, although he had to endure worse persecutions, and although he was not willed, even if he had been able to do it, to retaliate: yet he did not suffer it with a passive non-resistance; he turned to the man who beat him and took him to account, saying: "If I have spoken evil, bear witness of the evil; but if well, why smitest thou me?"

The doctrine "Render not evil for evil" is addressed to every single person as an individual. But it does not refer to the government, nor to the magistrate. If you are a judge and called upon to pronounce a verdict, the word has no reference to your judgment. We as persons have to renounce all egotism and all vindictiveness. For egotism and the ill-will of the human heart are the roots of all evil. Our egotism and the evil wants of petty personal desires must be renounced once for all and without reserve, not only where we do wrong, but also where we suffer wrong.

That Christ did not intend to teach the weak morals of non-resistance can be learned from his own demeanor. When he and his disciples came to Jerusalem, "Jesus went into the temple, and began to cast out them that sold and bought in the temple, and overthrew the tables of the moneychangers, and the seats of them that sold doves; and would not suffer that any man should carry any vessel through the temple. And he taught, saying unto them, Is it not written, My house shall be called of all nations the house of prayer? but ye have made it a den of thieves."

Christ did not render evil for evil where his personal interests were involved, yet if punishment is to be called an evil, he did not hesitate to render evil for evil in that dominion where he considered himself as the representative of Him that—according to his ideals of religious life—he felt had sent him.

Humanity, Christian and non-Christian, is under the influence of the sheep allegory still. One of the greatest biologists denies the existence of moral facts in nature, because the sheep and the deer are eaten by the wolves, and because in human society the same struggle for existence as in brute creation is fiercely fought, although with more refined weapons. The struggle for existence will continue, it can not be abolished, because it is a natural law, and sheepishness will never triumph in the world of real life. Having proved this, the scientist is satisfied, that nature is immoral.

Let us beware of the ethics of ovine morality. Morality is not negative, it is not mere submission to evil, no pure passivity, no suffering, simply: morality is positive. Not by the omission of certain things do we do right, but by straining all the faculties of mind and body to do our best in the struggle for life which we have to fight. We may be weak, and we may feel our weakness. The greater should our efforts be, to fight the struggle ethically. We may be poor in spirit and we may feel our want, but nature will supply us with that which we want, if we but earnestly struggle to acquire it. He who is strong in spirit and in body, he who feels his strength and misuses it, will not be the conqueror in the end. It is not the self-sufficient that are blessed; but those who are aware of their insufficiency. This only, in my opinion, can be the meaning when Christ says:

"Blessed are the poor in spirit: for theirs is the kingdom of heaven."

We must be on our guard against unfeeling sternness, yet on the other hand let us not drop into the other extreme. We must be on our guard against ethical sentimentality also. There is too much preaching about the sweetness of religion and the rapturous delight of ethics. Yet this saccharine religiosity is just as impotent and useless as that ovine morality which glorifies in its weakness and does not struggle for strength.

Austere rigidity in religion and ethics is like a rose without odor, it is life without gladness, and obedience without loving devotion. The passivity of a lamb-like submission is idealized weakness fortified and strengthened by moral vanity and sugared over with sentimental enthusiasm.

Religion and ethics, we do not deny, are full of sweetness and noble joys, yet at the same time they are stern; they are of an unrelenting severity and majesty. It is only the unison of both, the strength of austerity and the fervor of sentiment, that makes morality wholesome, sound, and healthy.

RELIGION AND ETHICS.

THERE are people who believe that theology and metaphysics have nothing to do with morality. Religious and philosophical world-conceptions, it is maintained, are one thing and ethical convictions are another. This is true in a limited sense only. It is true that the side issues of theology and metaphysics, which by theologians and metaphysical thinkers are generally considered as the most important of things, have as a rule little or no bearing whatever on morality. In so far, however, as Theology and Metaphysics discuss vital religious and philosophical problems, they *have* a certain relation to morality.

Morality depends on a sound conception of ourselves in relation to the world and, therefore, philosophical and religious errors will have an injurious effect upon morality.

If we allow ourselves to be carried away by impulse, we are not moral. Animals are un-moral. Their brutish conduct is not immoral; it is natural in them, as it becomes their brutish nature; and their good conduct (self-sacrifice of mothers for their young, etc.), although we justly praise it, cannot be properly considered as moral, because it is the result of instinct done from impulse and not an act of conscious deliberation. Man is moral in so far as he consciously and deliberately regulates his actions according to his relations to the All. Religion supplies him with the reason *why* the principles of his actions should be such as they are, and *why* he should do what he considers to be right and proper to do.

Religion, if understood to be our recognition of the Unity in Nature, teaches us to consider ourselves as parts of the whole; and who can doubt its strong influence upon all our conduct? The laws of the Universe govern also the motions of our body. Heat and gravitation operate as much in the functions of our organs as in the solar systems of the universe. Our lives depend upon surrounding nature, upon the atmosphere we breathe, the soil upon which we stand and the food which mother earth produces for us. Our existence is a continuous exchange and intercommunication with the whole "in which we live and move and have our being." The very pressure of the air upon our limbs is part of our life, which, if taken away, would cause instant dissolution.

But we are not only physical parts of Nature, we belong also to a higher order of natural growth which discloses ethical ideas and moral duties. The threads of each individual life are connected with the lives of other beings like ourselves, of beings whose origin is the same as ours and with whom we form one great family. These relations, although woven of invisible threads, are of no less importance than the coarser relations of our body to physical Nature. These relations of social and family life, if recognised, will teach us duties, and the performance of these duties is morality.

Religion, Science, Philosophy, Ethics and Morals accordingly are closely related to each other; religion is the recognition of the Unity in Nature which makes us feel that we are parts of it; Science is the study of the several departments of nature by observation and classification of its phenomena; philosophy is the result of the sciences, systematised. Ethics is the science of morals, and Morality is our behaviour regulated by religion, viz., by the recognition of the Unity of Nature in all its phases, the lower physical, the physiological and above all the social relations between man and man.

Those who are moral, prove that they *have* religion, for the moral man regulates his actions in accordance with his duties as implied by his relations to the All, especially to his fellow-beings. It is of great consequence to *have* religion in this sense, but it is of little consequence to *confess a* religion. *Religion* has to do with morals and morality, but all the different *religions*, i. e. the rites of churches, synagogues, and mosques, the various confessions, church-membership, etc., have little or no connection with morality, and if they have, it is only in so far as they contain *religion*.

False religions and wrong philosophies had always detrimental effects upon their adherents. The quietism of India has nipped in the bud a grand and rich civilisation, and the dualism of the middle ages has dragged many thousand victims to a shameful death for the alleged crime of witchcraft. The evil consequences of fundamental errors in philosophy and in religion bear witness to the dependence of morality on philosophy and religion. If you poison the religious or philosophical views of a man or of a nation, you will poison their morality also. The roots of man's intellectual life (viz., of that in man which makes of him a human being) are his convictions and his emotional inclinations (i. e., his philosophy and his religion), while his actions are the fruits thereof, by which we may recognise their soundness and vitality.

———————

THE ETHICS OF LITERARY DISCUSSION.

The ethics of literary discussion can be expressed in one sentence: Let the search for truth be your supreme maxim to which all other interests must be subordinate and subservient. Controversies which (not unlike duels) are waged for mere personal matters, have either to conform to this ethical maxim, or if they do not, they will be recognised as downright unethical or at least non-ethical.

The following rules are derived from the ethical maxim of literary discussion:

Never defend an opinion which you do not believe yourself. Never accept a belief which is not demonstrable. You must not only be convinced that it is so, but your arguments must be strong enough to convince impartial readers.

Strength of argument rests on the following conditions:

1. The facts upon which it is based, must be well established.

2. These facts must cover the whole field, so as to be exhaustive as instances.

3. The reasoning must be logical.

4. The presentation of the argument must be lucid.

5. Your presentation cannot be lucid if you are not

clear yourself. Accordingly, you must be ready to define every word you use.

6. Technical terms should not be employed unless their definitions are given.

7. Be careful that your words and especially your terms are used as they are commonly understood and not in a double or ambiguous sense.

8. Make the main points prominent and do not lose yourself in matters of detail, however interesting those details may be. They draw the attention of your readers and of yourself from the main subject.

These rules being observed, you can fearlessly await the most powerful adversary.

Before attacking the position of your adversary, try to understand his arguments from his standpoint.

Acknowledge fully where your adversary is right.

Where he uses an ambiguous term, state plainly in what sense the term would be allowable.

This is a matter of justice due to your adversary. To show justice in this way is advantageous first, to your opponent, and then, perhaps in a higher degree, to yourself, and what is most important, to the problem under discussion. It clears the situation and you thus limit the field of controversy to those points where you know your adversary to be wrong.

The points of agreement have become neutral ground which, it is true, your adversary can use for an honest retreat, if he chooses. However, his annihilation is not the object of the discussion, but the elucidation of truth. If he does not choose the chance of an honest retreat, his defeat will be the more inevitable, the more carefully the field of contest has been limited to his errors.

The weakness of an opponent is generally supposed to be the strength of his antagonist. This is utterly false. It must be a poor cause you defend, if it profits by the weakness of its adversaries. The strength of an adversary adds to your own strength if you defend a cause that is worth defending. The weakness of an adversary lowers you down to his own intellectual weakness. Therefore, do not have any discussion with weak opponents, and if you cannot avoid an encounter, do not take advantage of their weakness. The common issue is lost sight of by abusing an adversary for his weakness, ignorance, or faults. Consequently, you being the stronger, the duty of helping and promoting your adversary devolves on you. This should be done without ado, simply by giving information.

If your adversary uses rude language or derogatory expressions, there is no need of following his example or of attempting to out do him. Either do not answer his rant at all, or if you cannot avoid giving an answer, ignore all personal disparagement and confine your comments to the cause at issue. If you adopt the railing method of your adversary, you lower yourself to his moral inferiority.

Never use sophisms.

Sophisms easily impose upon large masses, but they do not delude the few independent thinkers who are perhaps silent by-standers. The ultimate result has never as yet depended upon the masses who judge rashly, but upon the judgment of the few independent thinkers who judge slowly but in most cases justly.

Sophisms are dangerous to the parties who employ them; sophisms will ultimately fall back and harm their own inventor. By using sophisms you venture on untenable ground, there to plant your colors, and if your enemy is on the alert, you will lose not only the position but your colors also. Sophisms afford incidental and transitory advantages.

If your adversary by negligence shows a hidden weakness or is guilty of a self-contradiction, point it out to him, stating at the same time how he should have expressed himself from his own standpoint. If his negligence is merely carelessness of verbal expression, you have settled the point for good. However, if the self-contradiction lies deeper, you have thus limited the field of discussion (as suggested above) to those points where the difference of the issues at stake will be seen to be primary and radical.

This always is the end toward which all honest and well directed discussion must tend. Even if the disputants cannot gain the best of one another, their discussion must elucidate the problem about which the discussion is waged. The disputants must learn by their discussion in how far they agree and wherein their differences consist : whether it is only a difference of words (which happens much oftener than is generally imagined), or a material difference. If it is a material difference, we must find out by the discussion, whether the difference is fundamental, i. e., whether the parties disagree because they start from different principles (which they have accepted as axioms) or whether it is a different interpretation of facts acknowledged by both parties, or whether one party takes its stand on facts which are not recognised by the other party as sufficiently established.

Whatever should be the result of a discussion conducted upon such ethical maxims, the discussion would never be entirely useless, but would be valuable in exact proportion to the issue at stake and the combined abilities of both opponents.

SEXUAL ETHICS.

SEXUAL ethics is the very core of all ethics. It is the most important sphere of human conduct, the tenderest, holiest, and most delicate realm of moral aspirations. When speaking of morality, we first of all think of sexual purity. So much is sexual ethics regarded as the very essence of morality! And no wonder that it is so. Consider but for a moment the importance of sexual relations! The future of our race depends upon them. The generations to come are shaped, they are created through sexual relations.

The legalized form of the sexual relation is called marriage. If marriage were not a sacrament, we ought to make it such, for it is the dearest, the most important, and most sacred of all human bonds.

The relation of parents to children is sacred indeed. It is the relation of the past to the present. Parents hand down the hallowed torch of spirit-life to the present generation; and if there is anything holier still, it certainly is the alliance between husband and wife to become parents and to devote themselves to the continuation of humanity and all the spiritual treasures of the race.

The sexual relation is a natural want produced through the necessity of self-preservation. The human soul yearns to live; it yearns to grow and to multiply. In the face of death it longs for immortality, but immortality is not granted to the individual and in order to become immortal an individual must grow beyond the limits of individuality. The natural consequence of these conditions is that immortality can spring from love only. Immortality must be gained by sacrifice, it must be taken by conquest, and there is but one power that can gain immortality. It is that power of which the Song of Songs says, "it is stronger than death." That one power is the holiness of the sexual relation, it is matrimonial love.

If we deprive sex-relation of its sanctity, it sinks down far below the most brutish acts of lowest animal life. Human sex-relation in which the spiritual elements of love and an exchange of soul are lacking degrades man and more so woman; it deprives them of their sanctity and sullies the holiest emotions they are capable of—the longing for immortal life. Animal sex-relations are at least natural. Animals yield to their natural wants without any consciousness of their importance or consequences. In the absence of thought, it is nature that acts in them. Immoral men and women, who prostitute the holiest sentiments because they imagine they find a pleasure in so doing, cease to remain natural and accustom themselves artificially to unnatural wants which weaken their bodies and poison their souls.

The apostle (in the Epistle to the Ephesians, vi. 2) speaks of the commandment "Honor thy father and mother," as being "the first commandment with promise." Reverence to parents is our willingness to receive the sacred torch of human soul-life with a grateful mind. Lack of reverence is a self-deprivation of this rich inheritance, and the highest reverence is shown not by a passive reception of merely conservative obedience, but by actively taking possession of the spiritual treasures by sifting them critically and by increasing their value. In fact, there is no passive receiving; all receiving is an active taking. Says Goethe:

> "What from your father's heritage is lent,
> Earn it anew to really possess it."

Greater than the promise of the fifth commandment is the blessing that accompanies sexual purity. Chastity is the condition of physical, mental, and moral health. When the Romans became acquainted with the valiant barbarians of the North, they recognized the natural holiness of the sexual relation as the source of their strength. Cæsar as well as Tacitus are fully aware of this fact and give in their historical accounts of German life with keen foresight due prominence to this most important factor in the evolution of a nation of barbarians.

The sexual instinct of man serves a most important and sacred purpose; it is the preservation of human soul-life, it is the attainment of immortality. If it is led into other channels, it decoys man into dangerous aberrations. Woe to those who find pleasure in depriving it of its sanctity! The curse that falls upon them will outlive their lives, for it will go down to their children and the children of their children.

It is not ethereal prudery that nature demands of us, not an extirpation or suppression of nature, but an elevation and purification, that the noblest features of nature's living and moving and being may be developed. A cynical attitude towards the mysteries of sexual life besmirches the soul of man with moral filth. Chastity has regard for laws that underlie the procreation of life, and reverence for the tenderest and most wonderful of nature's secret dispensations.

MONOGAMY AND FREE LOVE.

If we understand by free love what the word literally means, an absence of all compulsion to love so that love is granted and received as a free gift, what can be better, nobler, and more natural than free love? Love must always be free—or it is not love. Accordingly, free love is a matter of course, which in its proper meaning no one can dispute. Yet if we understand by free love that which as a rule is preached by most of the so-called apostles of free love, it would mean the absence of all restraint in the relation of the sexes, the destruction of its ideal element and the reign of licentious laxity. In that case it is only a beautiful name that has been given to an ugly monster; it is a devil that appears in the garment of an angel; it is moral filth praised as celestial manna.

There are laws of life which we must obey under penalty of perdition, and there are laws of love which we must obey under penalty of destroying the holiness of love or even defeating its end and purpose.

The purpose of love, that is of sexual love, is not the gratification of the sexual instinct, nor is it any pleasure that man may derive from such gratification. Wherever there is a gratification in love or in friendship, it is, regarded from the moral point of view, incidental; it is of secondary consideration and we need not speak of it here. The purpose of sexual love, its end and its holy law, is the welding of two souls into one so that a new soul-life may spring from it in which the two souls are inseparably fused.

What is soul? The Saxon poet says:

"Soul is form and doth the body make."

Soul is the form of a living organism. A fusion of souls actually takes place in the procreation of a new life; and this fusion of souls is one of those mysteries of nature which even, though science should succeed in explaining to our satisfaction its mechanical process, will forever remain a wonder before which we stand spell-bound in awe and admiration—a wonder which is grander and more miraculous than all miracles in which many of us are so fond of believing.

* * *

What is the law of love that must be obeyed? The law of love is obedience to the purpose of love, and the purpose of love is one of the holiest duties of man; it is the building up of our race. And this can be accomplished only if it is done with truthfulness, devotion, and self sacrifice.

The love of friendship between congenial minds, the love of the teacher to his pupils, of the preacher to his congregation, are also a building up, a preservation and a transference of soul-life in the human race; but conjugal love is devoted to the procreation of new souls, and without the sex relation of conjugal love humanity would die out.

Conjugal love in its legal form is called marriage, and the present form of marriage among all the civilized races is monogamy. Humanity has found by experience that society prospers best where the sexual relations are so arranged that one husband and one wife constitute the foundation of a family. The races in which polyandry prevails are rare exceptions; and wherever polyandry is the normal state of society, there is, as a matter of fact, no civilization, no culture, no progress. We have reasons to believe that polyandric tribes are a very low phase of human society, perhaps even a state of degeneration which in the end will lead to extinction.

Polygamy is practiced still in Asia, and it has been practiced among highly civilized people. Yet wherever monogamous and polygamous nations were rivals for supremacy, the monogamous nation proved always victorious in every kind of competition, in war as well as in peace.

There can be no doubt that monogamy is that form of matrimonial relations which best attains the ends of sexual love. Polygamous nations may have, but as a rule they do not have more children than monogamous nations, yet the children raised in monogamous family life are sturdier, healthier, and better educated. The institution of polygamy, while it degrades woman, easily induces man to marry merely for the gratification of their sexual appetites, and the seriousness of the duties of marriage is overlooked.

The ultimate purpose of marriage is the preservation of human soul-life, and if monogamy is more efficient in this one point than polygamy, if it enables man to raise a generation that loves freedom and delights in progress, it must be preferred whatever other advantages or pleasures might be connected with any other system of regulating the sexual relations in human society.

Monogamous nations are distinguished by love of freedom and by a progressive spirit; polygamous people are on the contrary easily enslaved. Their life as a rule shows a state of stagnancy, and their history consists of a series of court intrigues and palace revolutions.

Monogamy has become a holy institution to the nations of Aryan speech, because their civilization rests upon monogamous family life. So long as the moral sense of a nation is vigorous, it will most severely resent whatever threatens to destroy the holiness of monogamous family life. Thus the apostles of free love when they attempt to attack and destroy monogamy will meet with almost unanimous resistance.

* * *

The theory of free love in the sense of unrestricted sensuality is sometimes claimed to be the natural state, while matrimony is denounced by the defenders of free love as unnatural. If that were so, all the institutions of civilization ought to be considered unnatural. Raw food would be natural and cooked food unnatural; to live like the monkeys of the Sunda Islands would be natural, while plowing, sowing, and harvesting would be unnatural. Indeed the claim that free love is the natural state has been made only by most immature minds, who are without knowledge of the historical growth of our institutions, who are not familiar with the evils of such former states of society as are supposed to be more natural.

The defenders of free love very often lack all personal experience of harmonious and healthy family life. Not infrequently they have sprung from a marriage of ill-mated parents and have been too deeply impressed with certain incidental evils developed in such cases by the monogamous system. It would be a rare exception indeed if a father or a mother would advocate for their children the theory of unrestrained sexual intercourse.

Free love might perhaps be the correct theory, if such institutions as marriage could be judged from the standpoint of single individuals. The sex relation however is of greater concern than mere individual interest; and the problems rising therefrom must be judged from the higher standpoint of the common welfare of society.

The nature of human society develops certain relations which are wanting in the lower stages of animal life; but they are nevertheless just as natural. Who would say the oak is less natural than the lichen, only because the oak represents a higher stage in the evolution of plant life? The oak however would become unnatural, it would be in a morbid state, if its organs would degenerate so as to fall back to the lower stages of plant life.

Let us beware lest in trying to be natural, we should degrade ourselves into habits which may be natural to animals but are most unnatural to human beings—not that the satisfaction of the animal wants of man is unworthy of his higher nature, but that the animal way of satisfying them must be condemned.

* * *

It cannot be concealed however that as high an ideal as monogamy is, it sometimes demands great sacrifices; and the social sentiment which by law as well as by public opinion enforces the institution of monogamy, will sometimes have its victims. Marriages in which a man and a woman who for some reason cannot agree, are joined together until death shall part them, will produce misery that changes life into hell. There are also cases in which for some reason or other a legalization of the bond that has joined two noble souls in sacred love, could not take place. There are several well known instances even among great thinkers and geniuses of literary fame. There are some cases that cannot be measured by the usual standard of morality. It is a fact that men and women whose fates led them into paths that were different from the prescribed forms of marital relations suffered greatly from public prejudice. We should in such cases remember how kindly Christ treated the woman that was found guilty. "He that is without sin among you," Christ said, and we understand that he here refers to the sins against our sexual ideal of morality, "let him first cast a stone at her."

The sexual instinct in man is a most powerful element of his soul-life. It is dangerous to rouse it and more dangerous still to suppress or eradicate it. The whole vigor of natural forces is hidden in it. Sexual love wherever it grows is a serious thing to deal with. If it cannot have its way in legitimate channels, it will like steam that is shut up, break its way through laws and customs in spite of prejudices and public condemnation.

Let us therefore beware on the one hand lest we fall into temptation, and on the other hand when we see the mote in the eye of our brother, lest our judgment be too severe. Those who are without sin, beware that they preserve the purity of their soul. He who according to the holy legend of the Christian gospel was above all temptation, abstained from throwing a stone. He said in his lordly dignity to the adulteress: "Go and sin no more."

MORALITY AND VIRTUE.

Morality is taught in our churches and in our schools; it is preached in our religious and liberal congregations. And yet there is a strong doubt in the minds of many whether obedience to moral prescripts will be a help to a man who wants to get on in life. We hear it again and again that the moral man is the stupid man, the dupe of the smart impostor, while the man of the world, the man of business and of success must use misrepresentations. Strict honesty is said to be impossible. We are told by men of learning and experience who are supposed to know the world, that "the two sayings 'Be virtuous and you will be happy' and 'Honesty is the best policy,' are very questionable." And it is claimed by many that if that kind of honesty which never misrepresents nor ever keeps back part of the truth, were practiced, it would be difficult to carry on business.*

This view of life according to which the utility of honesty is of a doubtful character, which induces us to incline toward trusting in dishonesty as a good policy, which makes trickery and the methods of misrepresentation appear as promoting our interests, is the worst error, the falsest conception of life and the most

* See the article "A Few Instances of Applied Ethics" in *The Open Court* No. 219.

dangerous superstition that can prevail, and woe to that community where it becomes prevalent.

The grocer who sells impure goods as pure, the merchant who inveigles people to buy by false labels will succeed in cheating the public time and again. But let us not be hasty in forming our opinion, that cheating is advantageous; we shall find that in the long run this man cannot prosper through misrepresentations. There is but one thing that will wear, that is truth, and truthfulness is the only good policy.

The man who intends to cheat must be very smart, very wide-awake and very active in order to succeed, and in the end he will find out that better and easier rewards are allotted to the industry and intelligence that are used in the service of straightforward and honest purposes.

Several curious counterfeits are exhibited under glass to the inspection of the public in the treasury of the United States at Washington, and among them are two bills, one of fifty the other of twenty dollars, both executed with brush and pen only and yet they are marvels of exactness, and it must have been very hard to discover that they were imitations. No wonder that they passed through several banks before they were detected. The man who made them was an artist and he must have spent on their fabrication many weeks of close work. For the same amount of similar artistic and painstaking labor he would have easily realised more than double the return of the value which these counterfeits bear on their faces.

Is there any character more instructive than Ephraim Jenkinson in Oliver Goldsmith's world-famous novel "The Vicar of Wakefield." How successful Jenkinson was in his calling as a trickster and a rogue! and yet to be caught but once in a hundred times is for a rogue sufficient to ruin him forever. The Vicar and Jenkinson meet in the prison, and when the Vicar, having recognised by his voice the man who cheated him out of his horse, expresses surprise at his youthful appearance, the man answered, "Sir, you are little acquainted with the world; I had at that time false hair, and have learned the art of counterfeiting every age from seventeen to seventy." Jenkinson indeed appears as a master of his trade, yet he adds with a sigh: "Ah! sir, had I but bestowed half the pains in learning a trade that I have in learning to be a scoundrel, I might have been a rich man at this day."

Jenkinson is too smart to be wise enough to follow the experience of millenniums, laid down in the moral rules, and he found this out when he had leisure enough to think of his life within the prison walls. He says on another occasion to the Vicar:

"Indeed I think, from my own experience, that the knowing one is the silliest fellow under the sun. I was thought cunning from my very childhood: when but seven years old, the ladies would say that I was a perfect little man; at fourteen I knew the world, cocked my hat, and loved the ladies; at twenty, though I was perfectly honest, yet every one thought me so cunning that not one would trust me. Thus I was at last obliged to turn sharper in my own defence, and have lived ever since, my head throbbing with schemes to deceive, and my heart palpitating with fears of detection. I used often to laugh at your honest, simple neighbor Flamborough, and one way or another generally cheated him once a year. Yet still the honest man went forward without suspicion and grew rich, while I still continued

"tricksy and cunning, and was poor, without the con-
"solation of being honest."

Only a very superficial experience leads us to the assumption that wickedness is a help in the world and that the unscrupulous have an advantage in life. And this is a sore temptation to those who believe that it is so. Says Asaph in the seventy-third psalm:

"But as for me, my feet were almost gone. My steps had well nigh slipped.

"For I was envious at the foolish, when I saw the prosperity of the wicked.

"They are not in trouble as other men; neither are they plagued like other men.

"Therefore pride encompasseth them about as a chain; violence covereth them as a garment.

"Their eyes stand out with fatness: they have more than their heart could wish."

But the prosperity of the wicked is mere appearance. It is the state of the world as things seem to be, when only isolated instances are considered. The wicked may succeed a hundred times, but in the end they are sure to fail, and if they fail they are done with forever. An honest man may fail a hundred times and yet he may rise again, for his hands are clean and his conscience is not weighed down by guilt. Asaph continues:

"Then I went into the sanctuary of God and I observed their end.

"Surely thou didst set them in slippery places. Thou castest them down in destruction.

"How are they brought into desolation as in a covenant, they are utterly consumed with terrors."

Honesty is after all the best policy and he who does not believe it will have to pay for it dearly in his life.

But let us not go too far in our trust in honesty as

well as in all negative morality. Honesty is not enough to make success in life; honesty is not as yet virtue, and obedience to the several injunctions of the "thou shalt not" conveys by no means an indisputable claim to prosperity. True virtue is active not passive, it is positive, not negative.

What is virtue?

Morality as the word is usually understood is merely a refraining from wrong-doing; it is the avoidance of all that which does harm to our neighbor, which injures society or retards the growth and evolution of mankind. However, morality in order to be all it can be, ought to be more; it ought to be virtue, and virtue is the practically applied ability to do some good work. Virtue is activity, it is doing and achieving. And what is the good work which stamps activity as virtue? Virtuous is that kind of work which enhances the growth and evolution of mankind, which helps society, which promotes the welfare of our neighbors as well as of ourselves.

Mark! virtue is not exclusively altruistic; it is not opposed to egotism. Virtue may be altruistic, but there are sometimes very egotistic people who possess great virtues. Their virtues may be employed first and even so far their intentions go exclusively in the service of egotism. Nevertheless, they will designedly or undesignedly enhance the progress of mankind, and therefore we have to consider their abilities, their methods of action, their manners of work as virtues.

There are men of great virtue who have conspicuous moral flaws and it is not uncommon to judge of great men according to the pedantic morality of the Sunday school ethics. The bad boy who plays truant possesses sometimes more positive virtue than the good boy who is pliable and obedient to his teachers. It is a narrow view of morality and indeed an actually wrong ethics that cavils at the heroes of mankind, pointing out and magnifying their peccadilloes in order to obliterate and forget their virtues. Goethe whose greatness has often been detracted by the smallness of such dwarfs as have the impudence to speak in the name of morality said of Napoleon, the great conqueror and legislator:

>" At last before the good Lord's throne
> At doomsday stood Napoleon.
> The devil had much fault to find
> With him as well as with his kind.
> His sins made up a lengthy list
> And on reading all did Satan insist.
> God the Father, may be it was God the Son,
> Or even perhaps the Holy Ghost—
> His mind was not at all composed—
> He answered the Devil and thus began:
> 'I know it, and don't you repeat it here;
> You speak like a German Professor, my dear.
> Still, if you *dare* to take him, well—
> Then, drag him with you down to hell.'"

Lack of positive virtue is often considered as moral. Lack of courage is taken for peacefulness, lack of strength is taken for gentility, lack of activity is taken for modesty. If moral people are deficient in energy and ability, do they not deserve to be beaten by the wicked who possess energy and ability? Says Goethe in a little poem :

>" The angels were fighting for the right,
> But they were beaten in every fight.
> Everything went topsy turvy
> For the devil was very nervy,
> He took the whole despite their prayer
> That God might help them in their despair.
> Says Logos, who since eternity
> Had clearly seen that so it must be,
> ' They should not care about being uncivil
> But try to fight like a real devil,

> To win the day, to struggle hard,
> And do their praying afterward.'
> The maxim needed no repeating
> And lo! the devil got his beating.
> T' was done and all the angels were glad—
> To be a devil is not so bad."

Let us not be pusillanimous in ethics. It is pusillanimity which produces squinting views of morality. The morality of the pedant, the exhortations of the Sunday-school teacher, and the ethics of professors and lectures are not always correct, and if they are not exactly incorrect they are often insufficient or merely negative. The opinion that morality is no good guidance in life, that honesty is not always the best policy, that the unscrupulous, the deceitful, the immoral have a better chance in the struggle for life rests either on an insufficient experience or an insufficient conception of what ethics means.

Let us not be shaken in our trust in truth. Truthfulness toward ourselves and others is the best policy, it is the only possible policy that will stand for any length of time. Trickery, misrepresentation, deceit, imply certain ruin. At the same time let us remember that negative morality is not sufficient, we must have or acquire positive virtues. The omission of sins is not as yet the fulfilment of the law, the ideal of moral perfection is infinitely greater, it consists in building up the future of mankind in noble thoughts and energetic works.

ARISTOCRATOMANIA.

Envy of the rich is a very common feeling among the poor. And why is it so common? Because the rich are more fortunate in possessing wordly goods to satisfy not only their needs, but also any unnecessary wants. They have the means of procuring for themselves whenever they please all sorts of pleasures which because they are expensive lie outside the reach of the poor.

It is true that the rich have the means to procure themselves pleasures in an extraordinarily higher degree than the poor; but if the poor imagine that for that reason they actually enjoy life and life's pleasures better than the poor, they are greatly mistaken.

This is true in several respects. First the zest of pleasures is lost, if they are procured without trouble. Pleasure cannot be bought, pleasure must be felt, and the capability of having pleasure depends upon subjective and not upon objective conditions. The man who does not work lessens his capability of enjoyment in the same degree as he ceases to be in need of recreations; and pleasure which is no recreation after serious toil, which is not the satisfaction of a want, soon ceases to be a real pleasure, it becomes flat, stale and unprofitable.

The rich, in order to remain healthy in his spirit, in his sentiments, in his recreations and wants, must live like the poor man—not like those who are wretched and destitute, but like those who work for a living. The rich, be they ever so rich, must, for the mere sake of their mental and moral health, continue to be active, and their activity must have an aim and purpose, it must be productive of some good, it must be work of some kind.

The pleasures of the poor are, as a rule, richer and deeper in color than those of a certain class of typically rich people—viz., such rich people who noticeably appear and wish to appear as rich among their less fortunate fellow creatures; and the reason of this difference lies deeper still than in a mere lack of exertion and wholesome activity on the part of the rich. One of the most irresistible temptations of the rich, it appears, is their eagerness to be distinguished from their fellowmen as a special class of men, a peculiar and a higher species of the human kind. This is a disease which may be called aristocratomania, and it is one of the most deplorable diseases, not uncommonly proving fatal to the existence of noble and great families.

Aristocratomania is a disease which erects a barrier between special classes of men, not because the one is actually better, wiser, more moral, or nobler in character than the other, but because the one can indulge in luxuries in which the other cannot.

The aristocratomaniac is no aristocrat in the etymological and good sense of the word. He is not a better man than the rest of mankind; he is worse, he is a degeneration. His soul instead of being enlarged and widened has shrunk, and in the measure as it has shrunk it has lost in human interest, sympathy, and love.

The aristocratomaniac is perhaps charitable, he is kind, but his charity and his kindness appear offensive as soon as they are properly analysed, for their main element is a superstitious condescension.

The state of aristocratomaniacs is ridiculous and pitiable. It is ridiculous because of the vanity of their pride; it is pitiable because of the shriveled condition of their souls. The punctilious observance of social formalities has taken the place of cordiality and truthfulness. The fashionable ceremonial of society life has become the highest rule of conduct, but the real sentiments which ought to underlie the forms of social intercourse are neglected and forgotten.

The highest object of the aristocratomaniac is to burn incense before the altar of his God—the Puny Self which is fed with flattery and vanity. No emotion is permitted which would conflict with this deity, for great is the Puny Self and he is almighty in the soul of the aristocratomaniac.

Whenever the aristocratomaniac has injured or has given offense to his fellowman, the little word: "I beg your pardon," which by natural impulse wells up in a human soul, remains unspoken because the great Puny Self sees in it a humiliation of his majesty.

Why is there so little warmth in the family life of aristocratomaniacs? Brothers and sisters among the poor help one another, they rejoice at their joys and bear their woes in common. Does wealth produce a chill in the atmosphere so as to freeze out all cordiality and goodwill? Does wealth beget dissatisfaction, envy, jealousy, ill-will among men? Is the old Nibelungen-saga true that a curse rests on gold which will lead its owner to perdition? Certainly it takes a strong character to be wealthy and to remain human, kind-natured and broad-minded. The dearest and most sacred affections are too easily suffocated among the thorns and thistles of worldly goods. Proud of their possession of worldly goods the higher goods of truly human feelings are lost. As the mother of Christ said to Elizabeth:

"God hath filled the hungry with good things and the rich he hath sent empty away."

There are several causes of aristocratomania, for it is a very complicated disease and its symptoms show themselves in different ways, but one cause appears to be its main source and this one cause is the lack of solidarity with the interests of aspiring, toiling and progressing mankind. That which kindles sympathies in the hearts of men are common labor, common sorrows, common wants and common hopes. There is nothing of that between the aristocratomaniac and his fellowmen. He has with other aristocratomaniacs common joys, common fancies and fashions, common comforts and a common pride. But these feelings do not kindle sympathies.

There is a peculiar and a manlike sympathy in the dog who drags the cart of his poor master and earns a living as his help mate, sharing his master's labor and bread. But there is no such amiability in the snarling pug who idles away his time in the lap of his idle mistress. He is egotistic, impertinent and dissatisfied. He has also become infected with aristocratomania, for dissatisfaction is one of the most telling symptoms of the disease. Says Goethe in describing the symptoms of aristocratomania:

> "They 're of a noble house, that's very clear
> Haughty and discontented they appear."

There are among the poor a class of people who either from lack of strength, because the burdens of life are heavier than they can bear, or from lack of courage and good will, because they do not intend to work for a living, become spiteful and bitter. This disease is in many respects similar to aristocratomania. The aristocratomaniac feels himself exempt from the common lot of mortals, the spiteful poor thinks that he also ought to be exempt. He has the predisposition of becoming an aristocratomaniac, and being hopelessly shut out from the class to which his instinct leads him, he dreams of rising above the crowd of common mortals with the help of the masses by preaching hatred and destruction. This is the Marrat type of the demagogue, vanity, egotism and ambition are but too often the motives of him who pretends to be a reformer, imitating Christ in his denunciations only but not in his charity, love and self-renunciation. One of the most prominent social agitators actually exposed his main spring of action in quoting Virgil's verse:

> "Flectere si nequeo superos Acheronta movebo.
> [Can I not bend the Gods, I'll stir the under world.]

Moral health cannot be found in the aristocratomaniac nor in the would-be aristocratomaniac, but in the patient and plodding worker, be he rich or poor. He who has risen in his imagination above mankind and the sorrows of mankind has cut himself loose from the tree of humanity. The fate of aristocratomaniac families as a rule is sealed. They are doomed. Life is activity and wherever life ceases to be activity, it dries up and withers away.

Is this perhaps the meaning of Christ when he said that

> "A rich man shall hardly enter into the kingdom of heaven.
> "It is easier for a camel to go through the eye of a needle than for a rich man to enter into the kingdom of God."

These passages are strong and what they teach should not remain unheeded. There are two lessons which they teach, one of warning and the other of comfort. The warning is for the rich not to erect a barrier between themselves and humanity, not to allow their souls to be shriveled by wealth and pride of class, for the poor, not to be blinded by the advantages of wealth; wealth is not happiness and does not convey happiness. The real contents of life, its meaning, its import and its worth cannot be expressed in dollars and cents. We have to create the actual values of life ourselves.

But there is in Christ's words about the rich also a solace. The solace is for those who live their lives in the sweat of their brows. Life's strength is labor and sorrow. Let us not expect a different fate and we shall the more easily be able to meet the duties of life and to conform to the unalterable laws of mental and moral growth.

Let us not lose time with complaints, but let us be like Horatio:

> " As one, in suffering all that suffers nothing,
> A man that Fortune's buffets and rewards
> Has ta'en with equal thanks."

Let us preserve the elasticity of our minds and if we have to drudge, if we are surrounded with difficulties and disappointments, we shall bear them gladly and grow the stronger through their resistance. It is said that the palm tree, if weighed down by some heavy stone grows the more stately and the more straight raising its crown above all the other trees which either do not experience any resistance, or if they did, would not have the strength to overcome its pressure.

SOCIALISM AND ANARCHISM.

Social reformers and the enthusiastic prophets of a new mankind tell us that when their dreams are realised a radical change will take place in the nature of man. The coming man will lose all the vicious features of the present man; universal happiness will reign all the world over and humanity will become a homogeneous mass either of independent sovereigns or of well adapted members of society. The former extreme is called anarchism, the latter socialism or nationalism; and the exponents of either view expect from the application of their panacea a cure for all social diseases and the institution of a millennium upon earth.

How vain are the endeavors to construct an ideal Utopia either of an individualistic or socialistic humanity! Does it not prove that sociology is still in its infancy? Instead of studying facts, we invent and propose schemes.

The mistake made by anarchists as well as by socialists is that individualism and socialism are treated as regulative principles while in reality they are not principles; they are the two factors of society. Neither of them can be made its sole principle of regulation. You might as well propose to regulate gravity on earth by making either the centrifugal or the centripetal force the supreme and only law, abolishing the one for the benefit of the other.

Individualism and socialism are factors and cannot be made principles. This means: Individualism is the natural aspiration of every being to be itself, it is the inborn tendency of every creature to grow and develop in agreement with its own nature. We might say that this endeavor is right, but it is more correct to say that it is a fact; it is natural and we can as little abolish it as we can decree by an act of legislature that fire shall cease to burn or that water shall cease to quench fire. Socialism on the other hand is a fact also. "I" am not alone in the world; there are my neighbors and my life is intimately interwoven with their lives. My helpfulness to them and their helpfulness to me constitute the properly human element of my soul and are perhaps ninety-nine one hundredths of my whole self. The more human society progresses, the more numerous and varied become the relations among the members of society, and the truth dawns upon us that no advantage accrues to an individual by the suppression of the individuality of his fellows. First he, in so doing, never succeeds for good, and secondly the mutual advantage will in the end always be greater to all concerned the more the factor of individualism in others remains respected. Human society as it naturally grows is the result of both factors, of individualism and of socialism.

The anarchist proposes to make individualism, and the nationalist to make socialism the main principle of regulation for society. Are not these one-eyed reformers utterly in the dark as to the nature of the social problem? The social problem demands an inquiry into the natural laws of the social growth in order to do voluntarily what according to the laws of nature must after all be the final outcome of evolution. By consciously and methodically adapting ourselves to the laws of nature, we shall save much waste, avoid great pains, and acquire the noble satisfaction that we have built upon a rock : and no innovation is possible except it be a gradual evolution from the present state and the result of the factors which are at present active.

Socialism and anarchism are the two extremes, and all social parties contain both principles in different proportions. The republicans and the democrats represent the same opposition of centripetal and centrifugal forces in their politics. Party platforms are exponents of the forces that manifest themselves in the growth of society. They may be either symptoms of special diseases or indicators of a wholesome reaction against special diseases. A movement may be needed now in the direction of anarchism and now in that of socialism. We may now want a regulation of certain affairs in which the public safety and interest are concerned : for instance, in giving licenses to physicians and druggists, in the supervision of banks, in railroad matters, etc., etc.; and then again we may want a greater freedom from government interference. The temporary needs as they are more or less felt will swell the one or the other party.

It would be a misfortune, however, if one of these partisan forces could rush to the extreme and realise the social or anarchical ideal before its opposite had been deeply rooted at the same time in the hearts of the people. Social institutions not based upon liberty, or government interference to the suppression of free competition, would be exactly as insupportable as anarchy among lawless people who have no regard for the rights of others. But there is no danger that either extreme would entirely disappear to leave the whole field to the other alone. The law of inertia holds good in the psychical and sociological world no less than in the physical.

As the present man is the man of the past only further developed, so the coming man will be the present man only wiser, nobler, purer. There is no chance for a radical change of the nature of man or of the constitution of society. However there is a chance and more than a chance, there is a fully justified hope and a rational faith that man will continue to progress. Nature's cruel work of incessantly lopping off the constantly new appearing vicious outgrowths of human life through the survival of the fittest, and by an extirpation of the unfit, will in the future be performed by man himself, from the start, as soon as he has discovered the conditions under which these outgrowths become impossible.

Human society will in the future be more anarchistic in the same measure as it will be more socialistic. Not that socialistic institutions or laws will through an external pressure abolish competition and impose upon the individual more socialistic relations; nor that the abolition of laws will restrict government interference so as to give more elbow-room to individual liberty. Individual liberty will increase at the same ratio as the social instincts of mutual justice will become more than at present a part of every individual man. This has been the law of social progress in the past, it has made the republican institutions of the present possible and this law will hold good for the future also. Anarchism could be realised only where the laws of justice were inscribed in the hearts of all men, so that every man were a law unto himself; and perfect socialism can be realised only where every individual's greatest joy consisted in the ambition to serve the community. The former would be a state of altruistic individualists and the latter one of individualistic altruists. Both states are ideals and both are represented by more or less consistent parties which for the attainment of the same aim propose opposite means. These parties are exponents of certain forces that manifest themselves in the growth of society. It is well to understand both ideals and to sympathise with both, although the one as much as the other may be equally impossible, for evolution is a constant and a simultaneous approxmation to both ideals.

LOOKING FORWARD.

Human progress depends upon the dreams of enthusiasts. The inventor, the discoverer, and the reformer are dreamers, who prophet-like see in their imagination things that other mortals know not of. Every one of such men might very well say: "I had a dream which was not all a dream." Their dreams become realities and many such dreams are commonplace facts to us now. Indeed civilization consists of such realized dreams. How useful are these dreams!

We call dreams which are not all dreams, ideals. Why is not every dream as useful as a genuine ideal? Because the stuff of which the ideal is made—I mean the genuine ideal only—is taken from the actual state of things as they exist in reality, and handled according to the laws of nature.

James Watt took iron and steel and steam, and made them act according to their nature. He combined certain realities. He applied natural laws, and lo! the combination of his thoughts revolutionized the world, and lifted all humanity upon a higher level than it had occupied before. The genuine ideal is a dream that genius shapes out of reality.

We have become reverent toward the dreamer because of the usefulness of certain dreams. Dreamers, it appears, command our respect even if they are but dreamers. A certain man once learned at school that our atmosphere exercises a constant pressure of fourteen pounds upon every square inch of our body—constituting a total pressure of about forty hundred-weights upon the surface of the skin of an average adult person. This man had a dream that he lived upon a planet without an atmosphere. People felt so free and easy, in the absence of all pressure, that they moved about like winged angels. He told his dream to his neighbors, he wrote it down and published it, and it is the one hundred and ninety-first or second edition that is now being sold. Humanity builds altars to the dreamer, because he is a dreamer; he had a vision.

Every man that works for the progress of the human race has and ought to have our sincerest sympathy. We, all of us, should know that society in many respects,—perhaps in most respects,—is not what it ought to be. We have abolished slavery, but the laborer is not as yet the free, and independent, and intelligent man he ought to be ; not as yet is the employer the humane, and intelligent, and well educated man he ought to be. The people perish from want of knowledge ; it is knowledge that will make the laborer free, it is knowledge that will make the employer humane. Knowledge, if it is knowledge at all, means an acquaintance with facts as they really are, with natural laws and sociological laws, which latter are just as much laws of nature as gravitation or other natural laws are. And it is truth only that can make us free.

There comes a dreamer who flatly proposes to abolish the law of gravitation. He explains in a marvelously lucid sketch that every man who falls and breaks his leg, falls only because of the law of gravitation. Things are heavy because matter gravitates toward the centre of the earth. All the troubles of transportation are inconveniences due to gravity. There is no misfortune or annoyance that has not its root in this vilest of all natural institutions—gravity. Come therefore and let us abolish gravitation!

A dreamer like that is called an idealist, and great respect is paid him by the unkowing many. It is difficult to state whether such a dreamer, and all those infatuated by his dream, are to be envied or to be pitied for their illusions.

Mr. Bellamy depicts a state of society where there is no competition. Competition is the struggle for life among peaceful human beings. It is the struggle for life that created man and human society and all progress of the human race. But then there is much misery that arises from the struggle for life. The lesson that life teaches is, in my opinion, the admonition to make the struggle for life more humane. Let us therefore educate the growing generation better than the former generations, let us adapt ourselves to nature, let us break down artificial barriers between man and man, that the struggle for life may become a fair and honest fight for progress, that our competition may be an honest endeavor to do better and more useful work. Let us be fair to our enemies and to our competitors, and we shall soon find out, that the abler they are, the stronger and fiercer their competition is, the better it will be for us. They help us to progress, they force us to progress, however much worry they cause, we would certainly not be better off without them.

Why should the relation between employer and

employee be that of a master to his slave? It is partly now, and let us hope that in the future it will always be, looked upon as the co-operation of a worker with his co-workers, in which the one bears the main risk and will get a proportionate share of the profits, if there are any, while the others earn their fixed wages. Why should we abolish the principle of free enterprise, which encourages thrift, and progress, and invention, because there are some imperfections in its application?

In certain branches co-operation may, and I believe it will, become more practical than it is to-day. Such co-operation will in each case have to be based upon the freewill and assent of every independent individual, but it cannot—even not by the vote of a majority—be imposed upon the whole nation. And if it could, it would not work. It would change all trades into industrial armies and a few bosses would have to run and regulate the whole co-operative business of the nation. It would transform our present life of free enterprise and competition into an enormous penitentiary, only very humanely instituted—supposing that all convicts would willingly submit to the rules of the institute. The imperial army as well as the imperial post office and railroad service of Germany are a partial realization of Nationalism.

We want more chances for labor, more elbow-room for the courageous, especially for the poor. It is true, we demand that the license of the unprincipled be checked, but we do not want the liberty of anybody to be curtailed, be he a millionaire or an unskilled navvy.

Mr. Bellamy proposes to abolish the struggle for life. He has told us in his little book all the advantages of the scheme, and they are many. We can dispense with all the tedious inventions of civilization; we need no more private property, no money, no rewards for industry. But with the evils of competition,—which has produced our civilization,—we shall abolish the most divine blessings: human freedom, independence, responsibility, and above all self-reliance.

We are confident that "the present order may be replaced by one distinctly nobler and more humane." But the new order of things cannot be established by the proposed panacea of Nationalism and the abolition of competition. The new order must grow and evolve out of the present state of things, not otherwise than our present civilization developed out of savagery. In the new order of things we hope all unnecessary struggle will be avoided; we shall have less waste and a minimum of friction; yet the law of competition will remain in a future and better state of society just as powerful as it ever was since time immemorial and as it is to-day.

Nature has not designed man to live for the mere enjoyment of life. Nature under penalty of degeneration sternly demands and enforces a constant progress through struggle and work and sacrifice. And those who devote themselves to the pursuit of happiness, will soon find that they are following an *ignis fatuus* that leads them astray into the imperviable marshes of perdition. If a social reformer promises a millennium of happiness, be on your guard, for in that case he is misleading you. Look at his schemes with a critical mind and you will see that his Utopia is a fool's paradise.

Mr. Bellamy's book and its popularity is one of the most ominous symptoms of our time. It is an outcry for the satisfaction of material wants and for pleasures; a hunger for *panem et circenses* to be provided by the government, by the nation. The average citizens of Rome during the Punic wars were by no means rich, but they possessed an indomitable love of independence, and the Republic at that time rested upon a sound foundation. But when the Romans cried for *panem et circenses*, Liberty died and Caesar appeared. Caesar gave them *panem et circenses*, and the price they had to pay for the trouble he took, is known in history. The people who want to be taken care of and catered to with bread and pleasures, have forfeited their claim to freedom.

"Looking Backward" proposes to abolish the social law of gravitation which indeed causes many troubles in life but which at the same time produced and still produces our civilization. Thus the book is truly a looking backward to the primeval state of barbarism.

Let us cease to dream the useless dreams of abolishing the laws of nature. Let us rather abolish the artificial barriers between the so-called higher and lower classes. Give the poorest a chance to acquire as good an education as the richest command. Facilitate the opportunities of labor so that the industrious need not go begging for work. Thus we shall break down the hindrances that prevent progress, and in adapting ourselves to the laws of nature we shall better be prepared for a true and useful Looking Forward.

WOMAN EMANCIPATION.

One of the most important and at the same time noblest of our present ideals is the emancipation of woman. Woman is the weaker sex, because nature has destined her strength to be sacrificed for the perpetuation of the race. Woman represents the future of humanity; the immortality of mankind is entrusted to her. The burdens of life are upon the whole so divided that man must struggle with the adversities of conditions, while woman must suffer all the throes and woes which are the price of the continuance of human existence. He is the more active fighter, the worker, the hero; she is the passive endurer, the toiler, the martyr. He has under these conditions grown strong, physically and intellectually; she has grown noble. The activity of each being shapes its organism and models its character. Thus the virtues of man became daring courage, concentration of thought, and enterprising energy; the virtues of woman became abnegation of self, patience, and purity of heart.

Woman, being the weaker sex, has been and to a great extent is still held in subjection to the power and jurisdiction of the stronger sex. It is true that among cultured people the rudeness of this relation, has disappeared. The husband has ceased to be the tyrant of the household. He respects the independence of his wife and prefers to have in her a loving comrade rather than a pliant slave. Nevertheless progress is slow. It is perhaps not so much oppression by single persons as by traditional habits that is still weighing heavily upon woman, retarding the final emancipation of her sex.

Prof. E. D. Cope has written an article on the economical relation between the sexes* in which he emphasises woman's dependence on the support and protection of man. Professor Cope explains satisfactorily the present state of society, but he leaves out of sight the question whether this present state has to continue forever. His article is a scholarly investigation of existent conditions, but he does not touch the problem whether this is the only possible natural state or a special phase in the development of human sex-relations. We believe that the present phase is to be followed by another phase securing to woman a better, nobler, and more dignified position.

It may be conceded, as a matter of historical statement, that in the struggle for life women had to depend upon men for protection and sustenance. Yet it must not be forgotten that men in their turn also had to depend upon women. What are men without mothers and wives? How helpless is an old widower, and in spite of his so-called liberty how poor is the life of an old bachelor.

Professor Cope does not overlook this point, yet he maintains that women as a rule cannot make a living; he maintains that whenever they do, it is an exception and this is the reason why they must look for sustenance and protection from the stronger sex.

* *The Monist*, No. 1, p. 38.

Granted that this has been so; also granted that many women had to marry for this sole reason, must we therefore conclude that this wretched state of things is to continue forever? It may be true that there was a time when serfdom was an unavoidable state for a certain class of people who in a state of liberty would not make a decent living for themselves; slavery perhaps was a greater blessing to them than to their masters. Would that be a reason for continuing slavery in a higher state of social conditions?

The woman question has originated through the very progress of civilisation. In order to make a living a human being has no longer to depend upon physical strength, but mostly upon mental capacities, nay, more so upon moral qualities. Sense of duty is more important than muscle power, and sometimes even than skill. The time has come that at least in many branches a well educated woman can do the same work as a man, and she is no more dependent upon man for sustenance and protection.

This fact will not alter the natural relation of sex. Our women will not cease to marry, to bear and to raise children. Yet it will alter their position in this relation. They will no longer marry for the mere sake of protection, but for love alone. They will then enter marriage on equal terms; and thus they will obtain a more dignified place in human society.

It cannot be denied that woman is different from man. The average man is superior in some respects, and the average woman is superior in other respects. Neither man nor woman is the perfect man. True humanity is not represented by either. True humanity consists in their union, and in the consequences of their union, namely in the family.

Woman's emancipation does not involve any detraction from man's rights or duties. Man will not suffer from it, on the contrary, he will profit. It will raise our family life upon a higher stage and man will be as much a gainer in this bargain as the slave-holder who can employ free labor easier and cheaper than keep slaves. As no one would wish to re-establish slavery now, so in a later period no man would ever care to have the old state recalled when women married mainly for the sake of sustenance and protection.

Let me add that woman emancipation is slowly but assuredly accomplished, not by acts of legislature, but by a natural growth which no conservatism can stop. Acts of legislature giving more liberty and chances of making a living to woman, will not be the cause, they will come in consequence of a true woman emancipation. There are many steps taken in a wrong direction. Efforts are wasted especially by some over-enthusiastic women in making women like men, instead of making men and women equal. These erroneous aspirations are injurious to the cause, yet after all they cannot ruin it. There is an ideal of a higher, more elevated and a better womanhood, and this ideal (although it is often misunderstood) will be accomplished without the destruction of the womanly in woman.

DO WE WANT A REVOLUTION?

"Do we want a Revolution?" is the subject of an article by Mr. Morrison I. Swift which appeared in No. 166 of *The Open Court*. The question is answered in the affirmative; Mr. Swift declares: We want a revolution.

Mr. Morrison I. Swift is a young man and full of earnest enthusiasm for social justice and the elevation of the poor. He makes himself the attorney of the oppressed and hurls his shafts of indignation against the oppressors. He appears as the prophet of revolutionary reform, indicting a number of rich men, "because," he says, "they make our lives hard and dull."

Their crime, he declares, consists in being "willing in the present hour of enlightenment to accept the colossal advantages their place in an irrational system gives them, to use these perfectly prodigious powers selfishly." Not the slightest proof is adduced for this wholesale indictment. The indiscrimination in his collection of several well known names proves that Mr. Swift does not clearly know himself what they are guilty of. Are they arraigned for selfishness? Some of them are very active for the public good. Are they arraigned for possessing wealth? While none among them is poor, not every one of them is so extraordinary rich as Mr. Swift seems to imagine. Nor does the plaintiff indicate what these criminals ought to do in order to escape the condemnation of selfishness. Perhaps he would repeat the demand of Christ: "Go and sell all that thou hast and give to the poor and thou shalt have treasure in heaven?"

Plaintiff is a philanthropist and he kindly urges in extenuation that the rich are "victims of the system like the rest, victims of a sorry state of human nature." The personal indictment of these men seems to rest on the fact that they do not use their power to overthrow the social order. And this appears to Mr. Swift as the one thing that is needed. Having realised that there are iniquities and sufferings he is determined to promote revolution, because "life would be dishonorable on any other terms."

Mr. Swift undoubtedly hopes for a better system, which he supposes will come after the breakdown of the present system. He may be a nationalist or an anarchist, I do not know; and it matters little. Yet it is certain that rash youth only can so wantonly, although with best and purest motives, clamor for a revolution. Putting the question to himself whether or not we must be revolutionists, Mr. Swift declares "it is easy to make his choice."

Does Mr. Swift know what a revolution is? A revolution is a breakdown of society. It is not a building up, it is a tearing down. If is not evolution, but it is dissolution.

A revolution is a great public calamity which falls equally heavy on the rich and on the poor. Perhaps it falls heavier upon the poor, because as a rule they have less education and are ignorant of the course of events. The facts of the French revolution speak loud enough. Are they now forgotten? To every rich man who was guillotined hundreds of poor met with the same fate, and thousands were actually starved to death.

A revolution is like a deluge that, the dam being broken, sweeps over a valley. The deluge will drown the rich as well as the poor. It will often happen that a rich man may be drowned as well as a poor man; but after all, the rich man if he be warned in time, has better chances to escape.

Who will profit by revolutions? Not the laborer, he will be starved; not the employer of labor, he will be ruined. There is one class of men that will profit. It is the sharper; he whose business flourishes while and because all the world is covered with misfortune. There are people who undertake to fish in muddy waters. These people are the only ones that are benefited by public disturbances, calamities, and revolutions.

Several months ago I discussed the eventuality of a revolution with a leading anarchist of Chicago. I do by no means agree with anarchism; nor did this anarchist agree with my views, but he most emphatically joined me in denouncing the superstition so prevalent among many would-be reformers, that revolution can bring any salvation to society. He said, "When I was young and rash, I believed in revolution and hoped for a revolution; I thought to arrive at a higher state of society by a bee line road. But since I have seen more of life, I have ceased to believe in physical force. I then believed that society could be pulled up by the roots and pitched over the fence, and a new social machine, contrary to that which is, put in its place. I now see, that society is a slow growth, and the best we can do is to remove those special privileges, empowering the few to rob the many. Revolution, it is true, cannot be condemned under any and all circumstances. Revolution is, like war, always an evil, but in exceptional cases, it may happen to be the lesser evil. Revolution becomes necessary as soon as evolution has become an absolute necessity. Yet even then its necessity must be deplored, because all violence, bloodshed, and wars debase the higher sentiments of the race, and destroy the sanctity of human life; the progress which comes through peace, though slow it be, is the most certain and enduring."

There is but one way of improving the condition of the laboring classes; that is by evolution. We must enforce a better position of the workers by legal means, not with the bullet, but with the ballot. The road is slower, but it leads by and by to the desired aim.

The bee line road of revolution will not bring us nearer to a realisation of our ideals. In order to reach a better state of society by the slow process of evolution, we must educate mankind up to it, we must teach them a higher morality and a respect for law.

What a terrible error it is to preach justice and recommend the overthrow not of this or that law only, but of all laws and of the whole order of society.

Society is not an artificial system that can be constructed with arbitrariness. Society is an organism and the laws of its development are similar to those of living creatures, of plants and of animals. You can promote the growth of a tree, by digging round its stem, by watering the roots and pruning the dead branches in its crown, nay, you may inoculate a tree so that indeed the thorns may be made to bear figs or grapes. But if you pull out the whole tree, you will have to begin quite anew, and it will take a long while until it has reached that state again in which it is now.

Incendiary speeches are cheap means for agitators to become popular with the uneducated among our laboring classes. Yet I hope to see the time when our laborers will hoot at the demagogue who attempts to excite them with preaching hatred and ill will.

Yet the incendiary speeches of demagogues should not be ignored by the rich. We should recommend them to the rich for a careful perusal. There is certainly something wrong in a state of society in which young men, enthusiastic for justice, openly clamor for a revolution.

We advise the rich as well as the poor to weigh carefully Mr. Swift's proposition, not because we agree with him in the justice of a revolution, or in the advisability of preparing and preaching a revolution; on the contrary, because we should consider a revolution as the greatest public calamity, the evil consequences of which cannot be all foreseen. The probability, in my mind, is that the final result of a great revolution in the United States, would be the downfall of the republic and the establishment of an empire. A revolution, so far as I can see, will bring us no liberty but serfdom.

It is a law of nature that if a nation cannot govern itself, a usurper will keep order in that nation, and every revolution in a republic is a sign that the citizens are not able in a peaceful way to administer their public affairs.

The rich therefore, should heed the cry of alarm. They should consider that a revolution becomes an inevitable necessity as soon as the discontent of the poor in a country has reached a certain height at which their yoke appears to them unbearable.

Our society is by no means free from grievances, although they have not yet reached their fill. We should beware of the very beginning and mind all the symptoms of dissatisfaction. The greater the patience of the oppressed proves to be, the more formidable will be the outbreak of their indignation.

It is not good to build barriers between man and man; as says the prophet Jeremiah: "Let not the wise man glory in his wisdom; neither let the mighty man glory in his might; let not the rich man glory in his riches." And the apostle Paul writes to Timothy: "Charge them that are rich in this world, that they be not high-minded nor trust in uncertain riches."

The duties of those that have great possessions are greater than the duties of the poor. The more power a man has, the more imperative is his obligation to be just in all his dealing with his neighbors. The citizens of a republic should not attempt to make a caste of wealth; and ought to abhor all oppression of the poor. The employer must show his own independence and his sense of independence by respecting the independence of his employees. When weighing the worth of a man, let us not consider the amount of his property but the manliness and honesty of his character.

Is there any sense in admiring the aristocratic habits which have become fashionable with so many of our wealthy families? Let us exercise, ourselves, and teach our children to exercise, simplicity. Let us honor the democratic principles which so well become the citizens of a republic, and the mere idea of a revolution will become a ridiculous bugbear. Says Robert Burns:

> "Then let us pray that come it may,—
> As come it will for a' that,—
> That sense and worth, o'er a' the earth,
> May bear the gree, and a' that.
> For a' that, and a' that,
> Its coming yet for a' that,—
> When man to man the warld' o'er,
> Shall brothers be for a' that."

THE AMERICAN IDEAL.*

The United States of North America is a nation without a name. Poets hail our country Columbia, and Europeans call us simply Americans. Yet these appellations are not, properly speaking, names. Attempts have been made to provide the nation with a name, yet so far all the attempts have proved failures.

We need not care about a name. When we need a name, it will be given us. Much more difficult would it be to give ideals to a nation; yet luckily, although we are a nation without a name, we are not a nation without ideals.

We have high and great ideals, although they are neglected and forgotten by many; and some of our most influential politicians treacherously trample them under foot. We can say without boasting that our ideals are the noblest, the broadest, the loftiest of any in the world.

Our ideals are sublime because they are humanitarian, and thus this great republic of the West has become a bulwark against the evil powers of inherited errors and false conservatism. So long as it shall re-

* This article first appeared in *America* of Chicago.

main faithful to the principles upon which its constitution is founded, this republic will be a promise and a hope for the progress of mankind.

There is a prejudice in Europe against the ideals of America. It is fashionable in the old countries to represent Europe as the continent of ideal aspirations while America is described as the land where the dollar is almighty. Germans most of all are apt to praise the fatherland as the home of the ideal while the new world is supposed to be the seat of realistic avarice and egotism.

This is neither fair nor true, for there are as many and as great sacrifices made for pure ideal ends on this side of the Atlantic as on the other side. We maintain that Europe is less ideal than America. If impartial statistics could be compiled of all the gifts and legacies made for the public benefit, for artistic, scientific, and religious purposes, the American figures would by far exceed those of all Europe. In Germany the government has to do everything. It has to build the churches, to endow the universities, to create industrial and art institutions. If the government would not do it, all ideal work would be neglected, science would have to go begging, and the church would either pass out of existence or remain for a long time in a most wretched and undignified position. This state of affairs is not at all due to a lack of idealism among the people of the old world, but is a consequence of the paternal care of the government. The government provides for the ideal wants of its subjects; so they get accustomed to being taken care of. There is scarcely anybody who considers it his duty to work for progress, except where he cannot help it, in his private business, in industrial and commercial lines.

Scarcely anybody thinks of making a sacrifice for art, science, or the general welfare, and science and general welfare are looked upon as the business of kings and magistrates.

We live in a republic and the ideals of republican institutions are a sacred inheritance from the founders of this nation. We are no subjects of a czar or emperor, for in a republic every citizen is a king; and the government is the employé of the citizens. The highest officer of our government, the president of the United States is proud, when leaving the White House, of having tried to be a faithful public servant promoting the general welfare according to his best ability.

It is true that we are far—very far, from having realised our ideals. Our politics are full of unworthy actions, and many things happen of which we are or should be ashamed that they are possible at all in the home of the brave and the free. It is true also that many of our laws, far from expressing a spirit of justice and goodwill towards all mankind, are dictated by greed and egotism; further it is true that national chauvinism and national vanity go so far as to make any, even the sincerest, criticism of our national faults odious. Nevertheless we have our ideals and our ideals may be characterised in the one word humanitarianism.

How many there are who believe in the beneficial influence of petty advantages, unfairly gained by giving up the higher standard of justice and right! How many there are who suppress the cosmopolitan spirit of our ideals and foster a narrow exclusiveness which they are pleased to call patriotism. Their sort of patriotism will never benefit our country but will work it serious injury.

Our fourth of July orators pronounce too many and too brazen flatteries upon our accomplishments, and speak too little about our duties, when they represent us as that nation upon the development of which the future fate of humanity depends. There is too much talk about our freedom, as if no liberty had existed before the declaration of independence. What a degradation of the characters of our ancestry! Was it not love of liberty that set the sails of the Mayflower, was it not love of liberty that drove so many exiles over the Atlantic. Did the love of liberty not pulsate in the hearts of all the nationalities that make up our nation? Were not the Saxons, the Teutons, the sons of Erin, the Swiss, the French, the Italians, jealous of their liberties? does not their history prove the pride they took in preserving their rights and securing the dignity of their manhood? Love of liberty fought the battle of the Teutoburg forest even before the Saxon separated from his German brothers to found the English nation. Love of liberty was described by Tacitus as the national trait of the barbarians of the North whose institutions and customs and language have with certain modifications devolved upon the present generation now living in America.

Let us not undervalue our forefathers for the sake of a local patriotism; let us fully recognise the truth that we have inherited the most valuable treasures of our national ideals from former ages. In thus understanding how our civic life is rooted in the farthest past, we shall at the same time look with confidence into the darkness of future eras. Our present state is but a stepping stone to the realisation of higher ideals, for the possible progress of mankind is infinite and our very shortcomings remind us of the work that is still to be done.

Let us cherish that kind of patriotism which takes pride in the humanitarian ideals of our nation.

With our humanitarian ideals we shall stand, and without them we shall fall. So long as our shores remain the place of refuge for the persecuted, so long as our banner appears as the star of hope to the oppressed, and so long as our politics, our customs, our principles rouse the sympathy of liberty-loving men, our nation will grow and prosper; the spirit of progress will find here its home and the human race will reach a higher stage of development than was ever attained upon earth.

This great aim, however, can be attained only by a strong faith in the rightfulness and final triumph of the ideal, by perseverance and earnest struggle ; by a holy zeal for justice in small as well as in great things; by intrepid maintenance of personal independence and freedom for every loyal citizen ; and by the rigid observance of all duties political and otherwise so that the electors cast their votes in honesty and the elected fill their offices with integrity.

Historical investigations proved that the golden age must not be sought in the past. May we not hope that it lies before us in the future? Without believing in a millennium upon earth, in a state of ideal perfection, or in a heaven of unmixed happiness, we yet confidently trust that we can successfully work for the realisation of the golden age in our beloved home on the western continent—where the conditions are such as to leave us only these two alternatives: either the uneducated classes (among whom we have to count some of our richest citizens) will with their ballots and their influence in politics ruin the country, or they will, perhaps after many dearly bought experiences, be educated up to a higher moral plane.

Let us work for the American Ideal and let us hope for the future.

INDEX.

Abolition of dogmas, 29.
Achilles, 119.
Adaptation and evolution, 37.
Adversary, Understand your, 257.
Æschylus, 117, 119.
Agitators, 302.
Agnostic, 214.
Agnostic, The orthodox and the, 204.
Agnosticism, Three attitudes of, 214, 215.
Agreement between, 23.
Ahriman and Ormuzd, 94.
Amelioration, 142.
America, Ideals of, 305, 306.
Amos, 235, 236.
Anarchism and Socialism, 283, 285, 286, 287.
Anarchy, 139.
Angels and ethics, The, 274, 275.
Antinomies, 113, 114.
Antisthenes, 80.
Aristocracy of the mind, 196.
Aristocratic habits, 303.
Aristocratomania, 277, 278, 279.
Armour, Mr., 228.
Asaph, 272.
Athanasius, 92.
Atheism, 90, 104, 112, 193.
Atheism and dogmatic theism, 92.
Authority and freethought, 192.
Authority, God—to regulate action, 79.

Backbone and the struggle for life, 244.
Bacon, 107.
Baer, von, 44.
Baldur, 72, 73.
Beam in the eye, the, 198.
Belief in immortality, 175, 176.
Bellamy, 290, 291, 292.
Bible, 75, 76, 77, 221, 223.

Bible and ethics, 222.
Birth, 14.
Birth and death, 159, 160.
Blind guides, 199.
Bruno, Giordano, 230.
Bryant, quotation from, 162.
Buddha, 123, 125, 126, 143, 146, 171.
Buddhism, 143, 146, 148, 150, 171.
Burial, 14.
Burns, Robt., 304.

Cæsar, 262, 293.
Calvin, 30.
Ceremonies, 33.
Changes, 23.
Chastity, 262.
Christ, 6, 7, 17, 21, 22, 25, 33, 60, 73, 105, 106, 111, 121, 143, 146, 149, 151, 172, 208, 223, 224, 225, 236, 245, 246, 247, 248, 279, 280, 299.
Christ and non-resistance, 249.
Christ and the adulteress, 268.
Christ on the rich, 281.
Christianity, 4, 15, 32, 49, 143, 146, 148, 150.
Christmas, 71, 74.
Churches, 203.
Classical fairy tales, 50.
Clifford, W. K., 180.
Cold and heat, 95.
Commandments, 19, 21.
Competition, 239, 241, 248, 265, 285, 286, 290, 292.
Complexity, 39, 41, 42.
Conception of the world, 16.
Conciliation between science and religion, 61.
Conquest of death, 144.
Conscience, 54, 55, 75, 77.
Conservation of matter and energy, 139.

Conservation of soul-life, 140.
Co-operation, 291.
Cope, Prof. E. D., 295.
Cosmos, 10, 159.
Cosmos and design, 89.
Counterfeits, 270.
Criticism, 51.

Darwin, 43, 208, 227.
David, 56, 130, 213, 223.
David's son, 156.
Death, 14, 145, 147.
Death and birth, 159, 160.
Death and love, 185, 186.
Death, Conquest of, 144, 155.
Death, Dread of, 158.
Death no finality, 180.
Death the wages of sin, 141.
Deceptions, 56.
Dependence, Woman's, 295.
Design, 84.
Design and cosmos, 89.
Devil, Talmud on the, 95.
Devotion to truth, 60.
Dilettanteism, 234.
Direction in evolution, 94.
Direction of evolution, 95, 98.
Discussion of ethics, 256.
Disintegration increasing, 38.
Disparagement, 258.
Divine, 42.
Divinity and nature, 247.
Dogmas, Abolition of, 29.
Dogmas and science, 59.
Dogmatism, 34.
Doubt and faith, 227, 229.
Doubt in truth, 9.
Dreams and ideals, 288.
Dreams and progress, 288.
Dualism and monism, 240.
Duality of truth, 58.
Duty and work, 150.
Duty of clergy, 13, 14.

Easter, 151.
Ecclesiasticism and liberalism, 205.
Egg a symbol of resurrection, The, 152.
Egg, The Easter, 151.
Ego, 146, 147, 167, 168, 171.
Ego, No immortality of the, 187.
Ego, Surrender of the, 172.

Ego, Surrender of—no annihilation, 172.
Egotism, 273.
Eleusis, 165.
Emancipation, woman's, 297.
Enemies, Our—our counterpart, 242.
Entheism, 97.
Epicurus on death, 159.
Errors, 9, 120, 209.
Errors and truth, 208.
Esquimaux, 169.
Eternal youth and immortality, 161.
Ethical law and knowledge, 81.
Ethical man, 220.
Ethical questions and immortality, 177.
Ethical religion, 3.
Ethics, 37.
Ethics and Bible, 222.
Ethics and immortality, 154.
Ethics and pain, 219.
Ethics, Discussion of, 256.
Ethics in other parts of the universe, 218.
Ethics of altruism wrong, 241.
Ethics of arithmetic, 81.
Ethics of evolution, 47.
Ethics of struggle, 243, 244.
Ethics of the churches, 204.
Ethics, The angels and, 274, 275.
Evil and good, 95.
Evil and negative magnitudes, 96.
Evil and pantheism, 94.
Evolution, 5, 39, 41, 43, 44.
Evolution and adaptation, 37.
Evolution and immortality, 176, 179.
Evolution and intelligence, 83.
Evolution, Ethics of, 47.
Evolution, Laws of, 37.
Evolution not a material and not a mechanical process, 39.
Evolution of life not mechanical, 40.
Evolution of truth, 33.

Factors, Principles and, 283, 284.
Facts, 13.
Fairyland, 20.
Fairy tales and truth, 48.
Faith, 156.
Faith and doubt, 227, 229.
Faith in truth, 182, 193.

Feeling, 45.
Feelings and mind, 101.
Feel the truth, 53.
Fiction and truth, 153.
Fool of the gospel, The, 197.
Form, The secrets of, 152.
Free love, 263, 266, 267.
Freethinkers, Religion wanted among, 226.
Free, Thought not, 190.
Freethought, 189.
Freethought and authority, 192.
Freethought, Heroes of, 230.
Freethought, The God of, 193.
Freytag, Gustav, 141.
Fulfillment, 6.
Fulfillment of the law, 17.
Future, Religion of the, 147.

Genesis, 43, 99.
Germany, The paternal government of, 306.
Ghosts, 137, 138, 139, 141, 145.
Ghost-immortality, 168.
Ghost-soul, 167, 168, 169, 170.
God, 19, 21, 23, 79, 80, 111, 116, 146, 210.
God and nature not identical, 93.
God a mind? Is, 102.
God as an inventor, 86.
God, authority to regulate action, 79.
God everywhere, 97.
Godfathers, 14.
God-idea, 41.
God immutability of order, 87.
God is reality, 102.
God is spirit, 106.
God, Jahveh, 183.
God, Manifestations of, 105.
God not a man, 91.
God of nature, 80.
God superhuman, 88.
God the standard of ethics, 100.
Goethe, 77, 117, 119, 129, 170, 175, 184, 211, 221, 262, 274, 280.
Gold, A curse rests on, 279.
Goldsmith, Oliver, 270.
Good and evil, 95.
Good and truth, 59.
Gospel, 9, 11, 12.

Gravitation, Proposition to abolish, 289, 290.
Growth of soul, 42.
Guide through life, 46.
Guyau, on irreligion, 1.

Hagen and Rüdeger, 243.
Happiness, 148.
Happiness, pursuit of, 121, 124
Hated, 9.
Heat and cold, 95.
Hedonism, 47.
Heine, Heinrich, 215.
Hercules, 118.
Heroes of freethought, 230.
Hesiod, 50.
Heterogeneous, 38.
Higher life, 3.
Historical facts, 15.
Historical religion, 17.
History of religious progress, 61.
Holy Ghost, 221.
Hosmer, 50.
Homogeneous, 38.
Honesty, Need of, 235.
Honesty, Utility of, 269.
Hosea, 82.
Humboldt, 221.
Hume, 127, 230.
Huxley, 53.
Huxley on morality, 250.

Ibsen, Henrik, 137, 138, 141.
Iconoclast, 19.
Ideal world, 46.
Ideals and dreams, 288.
Ideals of America, 305, 306.
Idolatry, 87.
Idols, 80.
Immortal, 147.
Immortal, Ideals are, 165.
Immortal life, 144.
Immortality, 131, 145, 146, 151, 153, 154, 159, 160, 167, 169, 170, 171, 173.
Immortality and eternal youth, 161.
Immortality and ethical questions, 177.
Immortality and ethics, 154.
Immortality and evolution, 176, 179
Immortality and sex-relations, 260, 261.

Immortality and spiritism, 166.
Immortality and woman, 294.
Immortality, Belief in, 175, 176.
Immortality, Continuance of life, 181.
Immortality, No—of the ego, 187.
Immutability of order, God, 87.
Indestructibility, 170.
Indifference, 64.
Individual a part of the whole, 20.
Individual, the, 178.
Individuality, 135, 136.
Individuality preserved, 165.
Inertia, Law of, 25.
Infidel, 23.
Infinite, 108, 109, 110, 111.
Infinitude and mind, 102, 103.
Ingersoll, Robert, 209.
Ingersoll on immortality, 186.
Intelligence a machine, 86.
Intelligence, a machine of, 85.
Intelligence analysed, 83.
Intelligence and evolution, 83.
Ironbeard, 19.
Irreligious age, 1.

Jahveh, 183.
Jeremiah, 303.
John the Baptist, 33, 73.
Joy, Religions of, 148.
Justice alone insufficient truth, 56.

Kant, 44, 172, 230.
Kingdom of God, 32, 33.
Kingdom of God, Truth is the, 34.
King of truth, 34.
Knowledge and ethical law, 81.

Lamarck, 44.
Lasalle, 280.
Lavoisier, 25.
Law, 139.
Laws of evolution, 37.
Lazarus, Emma, 215.
Leckey on religion, 1.
Leo X., 231.
Lessing, 230.
Letter, 25.
Liberalism and ecclesiasticism, 205.
Life, 37.
Logos, 105, 106.
Love and death, 185, 186.

Luther, 31, 61, 144, 145, 164, 207, 209, 225, 230, 231, 232.

Macauley on the Puritans, 200.
Machine of intelligence, a, 85.
Man and woman, 296.
Man, Ethical, 220.
Manifestations of God, 105.
Marriage, 260, 264.
Matrimony, 14.
Maya, 122, 123, 171, 173.
Mayflower, 201, 308.
Mediums, 169.
Mene tekel, 235.
Metaphysical, 108.
Metaphysics and morality, 252.
Migrations of souls, 180.
Millennium, 292, 309.
Mind and infinitude, 102, 103.
Mind a world in ideas, 101.
Miracles, 19.
Mistletoe, 72.
Monism and dualism, 240.
Monogamy, 264, 265, 266.
Moral and religious faith, 15.
Morality, 131.
Morality active and passive, 273.
Morality and metaphysics, 252.
Morality and religion, 254.
Morality and success, 233.
Morality and theology, 252.
Morality and truth, 66.
Morality, Need of, 236.
Morality, positive, 274, 275.
Moses, 43.
Müller, F. Max, 108, 110, 112.
Mustard-seed, 156.
Mutability, 68.
Mutability and personality, 85.
Mysticism and truth, 52.
Mythology, 34, 66, 143, 150, 153.
Mythology of religion, 20.
Mythology, Truth in, 120.

Napoleon, 274.
Nation, 58.
Nationalism, 291, 292.
Natural laws of social growth, 285.
Natural selection, 38.
Nature, 75.
Nature and divinity, 247.

Nature and God not identical, 93.
Nature immortal, 2.
New, 1.
New ethics, 5.
New religion, 3, 12, 26.
Nibelungen saga, 243, 279.
Nirvana, 121, 123, 125, 144, 171.
Non-resistance, 245.
Non-risistance and Christ, 249.
Non-resistance, Struggle and, 248.

Observation, 33.
Omniscience, 104.
Open Court, The, 58, 60, 61.
Order, 10.
Order and struggle, 239.
Ormuzd and Ahriman, 94.
Orthodox, 19, 23, 33.
Orthodox and agnostic, The, 204.
Orthodoxy of science, 207.
Ovine Morality, 250, 251.

Paganism, 104, 105.
Paganism and personal God, 88.
Pain and ethics, 219.
Pain, diminution of, 218.
Palm tree, The, 282.
Panem et circenses, 293.
Pantheism, 90.
Pantheism, and evil, 94.
Pantheism, The blind side of, 93.
Parables, 66.
Parables, Meaning in, 120.
Paralogism, 167.
Paralogisms, 113.
Past, 6.
Pastor, 13.
Paul, 147, 164, 225.
Personal God and paganism, 88.
Personal intelligence laws, 84.
Personalities, Variety of, 135.
Personality, 88.
Personality and mutability, 85.
Phlogiston, 25.
Pilate, 33.
Pilgrims, The, 204.
Pious fraud, 176.
Pleasure, 26.
Pleasure, Craving for, 140.
Pleasures, The zest of, 276.
Pledged to truth, 12.

Polygamy, 265.
Poor and rich, 276-279, 299, 300, 302.
Positivism, 108.
Positivism and theism, 109.
Power and truth, 101.
Power, Religion ceased to be, 1.
Practical, 1.
Preservation of personality, 181.
Preservation of the individual soul, 178.
Principles and factors, 283, 284.
Problem, 216.
Progress, 35, 36, 37, 42, 244, 248.
Progress and dreams, 288.
Progress and truth, 194.
Prometheus, 117, 118, 119.
Pseudo-wisdom, 55.
Puritans, 164, 200, 202.
Pursuit of happiness, 121, 124.

Questions, 213, 214, 216.

Readjustment, 18.
Realistic, 1, 3.
Reality and the ideal, 288.
Reformation, 145.
Reformers, Retrogressive, 37.
Regulate conduct, 16.
Religion, 10, 35, 106, 112, 155, 156, 157, 200, 253.
Religion a fairy-tale, 49.
Religion and morality, 254.
Religion and power, 202.
Religion and scienc, 129, 206, 210, 211.
Religion a popularised system of ethics, 26.
Religion ceased to be power, 1.
Religion, Historical, 17.
Religion, Object of, 130.
Religion of facts, 66.
Religion of freethought, 189.
Religion of the future, 147.
Religion of joy, 148.
Religion wanted among freethinkers, 226.
Religion? What is, 60.
Religious and moral faith, 15.
Religious nature of evolution, 44.
Religious revolution, 4.
Republic, 307.

Republic and caste, 303.
Resignation, 143, 147.
Resurrection, 151.
Resurrection of the body, 182.
Resurrection, The egg a symbol of, 152.
Retrogressive adaptation, 38.
Retrogressive reformers, 37.
Revelation, 75.
Revelation and truth, 76.
Revelations, 76.
Reverence, 30.
Reverence for the merits, 6.
Revision, 28.
Revolution, 5, 298, 299, 300, 301, 302, 303, 304.
Rich and poor, 276–279, 299, 300, 302.
Right and truth, 47.
Right direction, 36.
Righteousness, 76.
Rudiger and Hagen, 243.

Saccharine religiosity, 251.
Sacrifices, 26.
Salter, 223, 225.
Salvation, 163.
Salvation of souls, 127.
Sausara, 121, 123.
Saviour, 74.
Scepticism, 227, 228.
Scherr, Johannes, 163, 164.
Schiller, 80, 93, 173, 210.
Schiller quotation, 69.
Schiller's Xenion, 188.
Schilling on revolution, George, 300, 307.
Schoolmen, 58.
Science, 5, 9, 15, 55, 106.
Science and dogmas, 59.
Science and religion, 129, 206, 210, 211.
Scriptures, 75, 76.
Search for truth, 112.
Selection, 38.
Selection in the struggle for existence, 237.
Self-preservation, Sexual relations, and, 260.
Sexual ethics, 260.
Sexual relations and self-preservation, 260.

Shakespeare, Quotation from, 170, 281.
Sheep, allegory, The, 250.
Sheep, The simile of a, 245.
Signs of the time, 2.
Simplicity, 42.
Sin, Death the wages of, 141.
Social growth, Natural laws of, 285.
Socialism and anarchism, 283, 285, 286, 287.
Socrates, 208.
Solace in death, 14.
Solomon's denial of immortality, 175.
Solovieff, 32.
Son of man, The, 119, 120.
Sophisms, 258.
Soul, 42, 128, 132.
Soul-life, Conservation of, 140.
Soul-life, Origin of, 40.
Soul of the Soul, The, 134.
Soul, Preservation of the individual, 178.
Soul, The human element of the, 284.
Soul, The unity of the, 133.
Souls, Hoarding up, 165.
Souls, Migration of, 180.
Souls of the past, The, 164.
Souls of the slain and the victor, The, 242.
Spencer, 38, 39, 46.
Spencer, Quotation from, 264.
Spinoza, 230, 247.
Spirit, 25.
Spirit, God is, 106.
Spiritism and immortality, 166.
Spiritism, its dearth of ideas, 169.
Spiritual life, 46.
Standard of ethics, God the, 100.
Strife, 240.
Strongest, 38.
Struggle and non-resistance, 248.
Struggle and order, 239.
Struggle and progress, 241.
Struggle, Ethics of, 243.
Struggle for existence, Selection in the, 237.
Struggle for life, 45, 291.
Struggle, the backbone of man, 244.
Struggle, The ethics of, 244.
Success, 4, 6.
Success and morality, 233.

Superhuman God, 88.
Supernatural, 19.
Superstition, 210.
Superstitions of science, 207.
Sursum, 44, 97.
Survival of the fittest, 237.
Survive, 38.
Swift, Morrison I., 298, 299, 302.

Talmud on the devil, 95.
Tat twam asi, 179.
Tauler, 173.
Tempter, the, 55.
Test of religion, 35.
Theism and positivism, 109.
Theism not wrong, 90.
Themis, 117.
Theology and morality, 252.
Thetis, 119.
Thinker a power, The, 63.
Thought, In the empire of, 25.
Thought not free, 190.
Tolerance, Misinterpretation of, 190, 191.
Tolstoï, Count, 246.
Treviranus, 44.
Trickery, 56, 233, 275.
Trust in truth, 192.
Truth, 8, 10, 11, 12, 42, 47, 56, 76, 78, 120, 206, 232.
Truth and controversies, 256.
Truth and errors, 208.
Truth and fairy-tales, 48.
Truth and fiction, 153.
Truth and freethought, 190.
Truth and good, 59.
Truth and morality, 66.
Truth and mysticism, 52.
Truth and power, 101.
Truth and progress, 194.
Truth and revelation, 76.
Truth and right, 47.
Truth appears to destroy, 182.
Truth, Devotion to, 60.
Truth in mythology, 120.
Truth is one, 20.

Truth is the kingdom of God, 34.
Truth objective, 191, 192.
Truth of the God-idea, The, 93.
Truth the supreme judge, 31.
Truth, Trust in, 275.
Truth, Unity of, 58.
Truth useful, 176.
Truth? What is, 66.
Truth will conquer, 65, 67.

Unity of the soul, The, 133.
Unity of truth, 58.
Universe, Is the—ethical, 217.
Unknowable, 40.
Useful, Truth, 176.
Utility of honesty, 269.
Utopia, 283.

Variety of personalities, 135.
Vicar of Wakefield, The, 270, 271.
Vice its curse, 139.
Victor, The souls of the slain and the 242.
Virgil, 280.
Virtue, 38, 273.
Voice crying in the wilderness, 12.

Watt, James, 288.
Weakness and morality, 245.
Weismann, 38.
Wheelbarrow, 224.
White House, 307.
Whole, 54.
Winnow the errors of the past, 92.
Wolff, Caspar Friedrich, 44.
Woman and immortality, 294.
Woman and man, 296.
Words religion, God and soul, 24.
Work and duty, 150.
World a cosmos, 41.
Worship of a personal God, The, 87.

Ygdrasil, 72.
Yule-tide, 71, 73.

Zeus, 118.

THE OPEN COURT

PUBLISHED EVERY THURSDAY BY

THE OPEN COURT PUBLISHING CO.

EDWARD C. HEGELER, Pres.   Dr. PAUI CARUS, Editor

P. O. DRAWER F.   169-175 La Salle Street.

CHICAGO, ILLINOIS.

The reader will find in *The Open Court* an earnest and, as we believe, a successful effort to conciliate Religion with Science. The work is done with due reverence for the past and with full confidence in a higher future.

The Open Court unites the deliberation and prudence of conservatism with the radicalism of undaunted progress. While the merits of the old creeds are fully appreciated, their errors are not overlooked. The ultimate consequences of the most radical thought are accepted, but care is taken to avoid the faults of a one sided view.

The Quintessence of Religion is shown to be a truth. It is a scientific truth which has been and will remain the basis of ethics. The Quintessence of Religion contains all that is good and true, elevating and comforting, in the old religions. Superstitious notions are recognised as mere accidental features, of which Religion can be purified without harm, to the properly religious spirit.

This idea is fearlessly and without reservation of any kind, presented in its various scientific aspects and in its deep significance to intellectual and emotional life. If fully grasped, it will be found to satisfy the yearnings of the heart as well as the requirements of the intellect.

Facts which seem to bear unfavorably on this solution of the religious problem are not shunned, but openly faced. Criticisms have been welcome, and will always receive due attention. The severest criticism, we trust, will serve only to elucidate the truth of the main idea propounded in *The Open Court*.

* * *

What is Science but "searching for the truth." What is Religion but "living the truth." Our knowledge of the truth, however, is relative and ad mits of a constant progress. As all life is evolution, so also Science and Religion are developing. With an enlarged experience of the human race they are growing more comprehensive, purer, and truer. Scientific truths become religious truths as soon as they become factors that regulate conduct.

The progress of Science during the last century, especially in the field of psychology, has produced the impression as if there were a conflict between Science and Religion, but there is no conflict and there cannot be any conflict between Science and Religion. There may be conflicts between erroneous views of Science as well as of Religion. But wherever such conflicts appear we may rest assured that there are errors somewhere, for Religion and Science are inseparable. Science is searching for truth and Religion is living the truth.

* * *

The Open Court pays special attention to psychology. Great progress has been made of late in a more accurate and scientific investigation of the human soul. While the new conception of the soul will materially alter some of the dogmatic views, it will not affect the properly religious spirit of religion, it will not alter the ethical truths of religion but will confirm them and place them upon a scientific foundation.

Since we have gained a scientific insight into the nature of the human soul, the situation is as thoroughly altered as our conception of the universe was in the times when the geocentric standpoint had to be abandoned. The new psychology which may briefly be called the abandonment of the ego-centric standpoint of the soul will influence the religious development of humanity in no less a degree than the new astronomy has done. At first sight the new truths seem appalling. However, a closer acquaintance with the modern solution of the problems of soul-life and especially the problem of immortality shows that, instead of destroying, it will purify religion.

The religion of *The Open Court* is neither exclusive nor sectarian, but liberal; it seeks to aid the efforts of all scientific and progressive people in the churches and out of them, toward greater knowledge of the world in which we live, and the moral and practical duties it requires.

ESPECIAL ATTENTION DEVOTED TO QUESTIONS OF ETHICS, ECONOMICS, AND SOCIOLOGY. The work of *The Open Court* has been very successful in this department. Discussion has been evoked on almost every topic treated of. Wheelbarrow's contributions to practical economics, Prof. E. D. Cope's and Moncure D. Conway's treatment of current sociological questions, Dr. G. M. Gould's, Mrs. Susan Channing's, and A. H. Heinemann's examination of criminal conditions and domestic relations, Gen. M. M. Trumbull's trenchant criticisms of certain ethical phases of our political life. The discussion between Wm. M. Salter, Profesor Jodl, and the Editor on THE ETHICAL PROBLEM, and many other contributions of note by Dr. S. V. Clevenger, Chas.

K. Whipple, J. C. F. Grumbine, George Julian Harney, John Burroughs, Wm. Schuyler, F. M. Holland, Ednah D. Cheney, E. P. Powell, Dr. Felix L. Oswald, Prof. Joseph Le Conte and others have been received with marked favor.

Authorised translations are made from the currant periodical literature of Continental Europe, and original contributions obtained from the most eminent investigators of England, France, and Germany.

In the philosophy of language may be mentioned the recent contributions of Max Müller on THE SCIENCE OF LANGUAGE, the translations from Noiré's works on THE ORIGIN OF LANGUAGE, and the essays of Mr. T. Bailey Saunders on THE ORIGIN OF REASON.

Articles on vital problems of PSYCHOLOGY and BIOLOGY, appeared from the pens of Th. Ribot, Alfred Binet, Ernst Haeckel, Prof. Ewald Hering, Prof. A. Weismann Prof. E. D. Cope, and others.

TERMS OF SUBSCRIPTION:

For One Year } Throughout the Postal Union	$2.00
For Six Months }	1.00
Australia, New Zealand, and Tasmania, One Year	2.50
Single Copies	5 Cents
Volume I	Bound, $4.00; Unbound, $3.25
Volume II	" 4 00; " 3.25
Volume III	" 3.00; " 2.25
Volume IV	" 3.00; " 2 25

Express charges, or postage, extra on back numbers.

THE OPEN COURT PUBLISHING CO.,

169-175 La Salle Street.   Post Office Drawer, F.

CHICAGO, ILLINOIS.

THE MONIST.

A QUARTERLY MAGAZINE
PUBLISHED BY
THE OPEN COURT PUBLISHING CO.

Editor: DR. PAUL CARUS.   *Associates:* { E. C. HEGELER, MARY CARUS.

PRICE, 50 CENTS. $2.00 PER YEAR.

THE MONIST is a magazine which counts among its contributors the most prominent thinkers of all civilised nations. There are American thinkers such as Joseph LeConte, Charles S. Peirce, E. D. Cope, Moncure D. Conway (the latter a native Englishman, but a resident citizen of the United States). There are English savants such as George J. Romanes, James Sully, B. Bosanquet, and the famous Oxford Professor, F. Max Müller. There are Germans such as Justice Albert Post, the founder of ethnological jurisprudence, Professors Ernst Mach and Friedrich Jodl, French and Belgian authors such as Dr. A. Binet and Professor Delbœuf. The Italians are represented by the great criminologist Cesare Lombroso and the Danes by their most prominent thinker Prof. Harald Höffding. Each number contains one or two letters on bibliographical and literary topics from French, German, or Italian scholars.

The international character of the magazine appears also in a rich review of English and foreign publications. Each number contains a synopsis of the most important books and periodicals, American as well as European, in the philosophical, ethical, psychological, and physiological fields.

THE MONIST represents that philosophical conception which is at present known by the name of "Monism." Monism, as it is represented in THE MONIST, is in a certain sense *not* a new philosophy, it does *not* come to revolutionise the world and overthrow the old foundations of science. On the contrary, it is the outcome and result of science in its maturest shape.

The term "Monism" is often used in the sense of one-substance-theory that either mind alone or matter alone exists. Such theories are better called Henism.

Monism is not "that doctrine" (as Webster has it) " which refers all phenomena to a single ultimate constituent or agent." Of such an "ultimate constituent or agent" we know nothing, and it will be difficult to state whether there is any sense in the meaning of the phrase "a single ultimate constituent or agent."

Monism is much simpler and less indefinite. Monism means that the whole of Reality, i. e. everything that is, constitutes one inseparable and indivisible entirety. Monism accordingly is a unitary conception of the world. It always bears in mind that our words are abstracts representing parts or features of the One and All, and not separate existences. Not only are matter and mind, soul and body abstracts, but also such scientific terms as atoms and molecules, and also religious terms such as God and world.

Our abstracts, if they are true, represent realities, i. e. parts, or features, or relations of the world, that are real, but they never represent things in themselves, absolute existences, for indeed there are no such things as absolute entities. The All being one interconnected whole, everything in it, every feature of it, every relation among its parts has sense and meaning and reality only if considered with reference to the whole. In this sense we say that monism is a view of the world as a unity.

The principle of Monism is the unification or systematisation of knowledge, i. e. of a description of facts. In other words: There is but one truth, two or several truths may represent different and even complementary aspects of the one and sole truth, but they can never come into contradiction. Wherever a contradiction between two statements appears, both of which are regarded as true, it is sure that there must be a mistake somewhere. The ideal of science remains a methodical and systematic unification of statements of facts, which shall be exhaustive, concise, and free from contradictions—in a word the ideal of science is MONISM.

Monism, as represented by THE MONIST, is a statement of facts, and in so far as it is a statement of facts, this Monism is to be called POSITIVISM. This Positivism however is different from Comtean Positivism, which latter would better be called agnosticism (see *The Monist*, Vol. ii, No. 1, p. 133-137). There is a mythology of science which is no less indispensable in the realm of investigation than it is in the province of religion, but we must not forget that it is a means only to an end, the ideal of scientific inquiry and of the monistic philosophy being and remaining a simple statement of facts.

Although the editorial management of THE MONIST takes a decided and well defined position with respect to the most important philosophical questions of the day, its pages are nevertheless not restricted to the presentation of any one special view or philosophy. On the contrary, they are open to contributors of divergent opinions and the most hostile world-conceptions, dualistic or otherwise, are not excluded.

PRESS NOTICES ON "THE MONIST."

"The establishment of a new philosophical quarterly which may prove a focus for all the agitation of thought that struggles to-day to illuminate the deepest problems with light from modern science, is an event worthy of particular notice."—*The Nation*, New York.

"The articles are of the highest grade."—*The Inter Ocean*, Chicago.

"No one who wishes to keep abreast of the most widely extended and boldly pushed forward line of philosophically considered science, can do better than attempt to master the profound yet lucid studies set forth in *The Monist*."—Ellis Thurtell, in *Agnostic Journal*.

"*The Monist* will compete most dangerously with the leading magazines of our own country.... *The Monist* is decidedly the morning star of religious liberalism and philosophical culture."—Amos Waters in *Watts's Literary Guide*, London.

".... demands and will repay the attention of philosophical inquirers and thinkers."—*Home Journal*, New York.

"It will take rank among the best publications of its class. We hope that it will receive the support to which its merits certainly entitle it."—*Evening Journal*, Chicago.

"It is both a solid and a handsome quarterly."—*Brooklyn Eagle*.

"The periodical is one of the best of the solid publications of the kind now before the public. The articles are substantial, clever, and catching in subject."—*Brighton Guardian*.

"It is a high-class periodical."—*Philadelphia Press*.

"One of the most solid serials of the times. All will be inclined to give a cordial welcome to this addition to scientific and philosophical literature."—*Manchester Examiner*.

"The articles are admirable."—*Glasgow Herald*.

"The subjects are treated with marked ability."—*Ulster Gazette*, Armaugh.

"A desideratum in the department of philosophical literature."—*Boston Transcript*.

"We welcome it to our homes and firesides."—*San Francisco Call*.

"Its merit is so exceptional that it is likely to gain a national, even a European recognition, before it has gained a local one. It deserves to be widely known."—*The Dial*, Chicago.

"We very heartily welcome this quarterly as a great help in the investigation of psychological questions."—*Boston Herald.*

"*The Open Court* and *The Monist* are unusually worthy of perusal by thinkers in the various departments of knowledge and research "—*Dubuque Trade Journal.*

"It is filled from cover to cover with choice reading matter by some of the most noted home and foreign metaphysical psychological thinkers and writers of the age."—*Medical Free Press,* Indianapolis.

"Every reader and investigator will find *The Monist* a most valuable and attractive periodical." *Milling World,* Buffalo.

"The reader will, by an attentive perusal of this most promising magazine, easily bring himself *au courant* with the best modern work on psychological and biological questions. The magazine deserves to take that established and authoritative position which we very cordially wish on its behalf."—*Literary World,* London.

"This magazine will be received with eagerness in the closet of many a student." *Hampshire Chronicle,* Winchester.

"*The Monist* is first-class, and numbers amongst its contributors the most eminent students of science and philosophy in England and America. There is no better journal of philosophy in England."—*Echo,* London.

"Those with a taste for "solid" reading will find their desire gratified here."—*Leicester Chronicle.*

"The October number of *The Monist* covers a wide area, and if it had no other claim upon popular favor than that of variety that in itself ought to be a sufficient guarantee to ensure it success. But it possesses the additional recommendation of being ably and brightly written."—*Morning News,* Belfast.

"The journal numbers amongst its contributors the most eminent students of science and philosophy in England and America."—*Sussex Advertiser.*

"In this number *The Monist* has sustained the high reputation of the three preceding issues. Two things are necessary to constitute a good quarterly, able contributors, and a live editor *The Monist* has both. The articles are all on living questions, practical as well as theoretical. If *The Monist* sustains the position already reached, it will be indispensable to every student who wishes to keep pace with current thought."—*The Canadian Methodist Quarterly.*

PUBLICATIONS
—OF THE—
OPEN COURT PUB. CO.,

169-175 LA SALLE STREET, CHICAGO, ILLINOIS.

THREE INTRODUCTORY LECTURES ON THE SCIENCE OF THOUGHT. By F. MAX-MÜLLER. (Sole Agents in England: Longmans, Green, & Co.)

1. The Simplicity of Language; 2. The Identity of Language and Thought; and 3. The Simplicity of Thought. Cloth, 75 Cents.

Prof. F. Max Müller sets forth his view of the identity of Language and Thought, which is a further development of Ludwig Noiré's theory that "man thinks because he speaks." Language is thought, no thought is possible without some symbols, be they spoken or written words.

THREE LECTURES ON THE SCIENCE OF LANGUAGE. By PROF. F. MAX MÜLLER.

With a Supplement "MY PREDECESSORS." Cloth, 75 Cents.

Prof. F. Max Müller points out that the difference between man and animal isdue to language, yet there is no mystery in language. He shows the origin of language as developed from the clamor concomitans of social beings engaged in common work. Thought is thicker than blood, and the bonds of the same language and the same ideas are stronger than of family and race relation.

THE PSYCHOLOGY OF ATTENTION. By TH. RIBOT. (Sole Agents in England: Longmans, Green, & Co.) Authorised Translation. Cloth, 75 Cents.

THE DISEASES OF PERSONALITY. By TH. RIBOT. Authorised translation. Cloth, 75 Cents.

The works of Th. Ribot explain the growth and nature of man's soul. The former book, "The Psychology of Attention" is an exposition of the mechanism of concentrating the will upon a special object, thus showing the cause of the unity of the soul and throwing light upon the nature of the ego. The latter book, "The Diseases of Personality" elucidates the hierarchical character of man's psychic life which rises from simple beginnings to a complex structure. The growth of personality is shown by an analysis of its diseases, and its evolution is set forth in the instances of its dissolution.

THE PSYCHIC LIFE OF MICRO-ORGAN-ISMS. By ALFRED BINET. (Sole Agents in England: Longmans, Green, & Co.) Authorised Translation. Cloth, 75 Cents.

A special fascination is attached to the wonders of the world of psychic life in a drop of water. It is astonishing how much these tiny creatures behave like ourselves!

ON DOUBLE CONSCIOUSNESS. New Studies in Experimental Psychology. By ALFRED BINET.
Price, 50 Cents.

Our conscious life is only part of our soul's existence. There are subconscious and even unconscious states and actions taking place in man which are of a psychic nature. M. A. Binet gives an account of his experiments in this field.

EPITOMES OF THREE SCIENCES.
1. COMPARATIVE PHILOLOGY. By PROF. H. OLDENBERG.
2. COMPARATIVE PSYCHOLOGY. By PROF. J. JASTROW.
3. OLD TESTAMENT HISTORY. By PROF. C. H. CORNILL. Cloth, 75 Cents.

The present state of our knowledge in these three sciences, so important for religious thought, is presented in this book by three specialists.

THE ETHICAL PROBLEM. By DR. PAUL CARUS.

Three Lectures Delivered at the Invitation of the Board of Trustees before the Society for Ethical Culture of Chicago, in June, 1890. Cloth, 50 Cents.

The Ethical Problem is a criticism of the position of the societies for ethical culture. They propose to preach ethics pure and simple without committing themselves to any world-conception of religion or philosophy. It is shown here that our views of morality always depend upon our view of life: every definition of good presupposes a certain world-conception.

THE SOUL OF MAN. An Investigation of the Facts of Physiological and Experimental Psychology. By DR. PAUL CARUS.

With 152 illustrative cuts and diagrams. 474 pp. Cloth, $3.00.

This book elucidates first the philosophical problem of mind, showing that mind is not motion but the subjective state of awareness accompanying certain motions of the brain. It describes the physiological facts of the nervous system and the experiments of hypnotism, and after a discussion of the Nature of Thought, Consciousne s, Pleasure, and Pain, it presents the ethical and religious conclusions derived from these considerations.

THE IDEA OF GOD. By Dr. Paul Carus.

A disquisition upon the development of the idea of God. Paper, 15 Cents.

The different conceptions of God (Polytheism, Monotheism, Pantheism, and Atheism) are discussed and God is defined as the moral law of the world which is recognised as the authority in accord with which we have to regulate our conduct. This view is called Entheism.

FUNDAMENTAL PROBLEMS. By Dr. Paul Carus. (Sole Agents in England: Longmans, Green, & Co.) Second Edition. Revised and Enlarged.

Cloth, $1.50.

Monistic Positivism, as presented in Fundamental Problems, starts from facts, and aims at a unitary conception of facts. Knowledge is a description of facts in mental symbols. Sensations are the data of experience yet the formal aspect of facts is recognised in its all-important significance. Agnosticism is rejected, and the ethical importance of a positive world-conception insisted upon. The appendix consists of a number of discussions in which the author considers all the objections made by critics of many different standpoints.

HOMILIES OF SCIENCE. By Dr. Paul Carus.

Gilt Top. Elegantly Bound. $1.50.

Short ethical exhortations and sermon-like discussions of religious, moral, social, and political topics made from a standpoint which might briefly be characterised The Religion of Science.

WHEELBARROW. ARTICLES AND DISCUSSIONS ON THE LABOR QUESTION.

Cloth, $1.00

This book is written by Gen. M. M. Trumbull and contains the very life-blood of his experiences. It is a collection of articles and discussions on the labor problem, and as the author has worked for many years as an unskilled laborer, he has a right to be heard and indeed his views are liberal as well as just and are nowhere lacking in a healthy moral spirit.

THE LOST MANUSCRIPT. A Novel. By Gustav Freytag. Authorised translation. Elegantly bound, $4.00. In one volume bound in cloth, good paper, $1.00.

The author writes as a motto for the American edition:

"A noble human life does not end on earth with death. It continues in the minds and the deeds of friends, as well as in the thoughts and the activity of the nation."

Gustav Freytag did not write his novel with the intention of teaching psychology or preaching ethics. But the impartial description of life does teach ethics, and every poet is a psychologist in the sense that he portrays human souls. This is pre-eminently true of Gustav Freytag and his novel "The Lost Manuscript."

www.ingramcontent.com/pod-product-compliance
Lightning Source LLC
Chambersburg PA
CBHW031856220426
43663CB00006B/643